The R.A.M.S. Library of Alchemy

Volume 27

The Vegetable Work
by
Johan Isaac Hollandus
(Johannes Isaaci Hollandi)

R.A.M.S. Publishing Company

The Vegetable Work

by

Johan Isaac Hollandus
(Johannes Isaaci Hollandi)

Produced by

Restorers of Alchemical Manuscripts Society
1978

R.A.M.S. Publishing Company

R.A.M.S. Publishing Company
117 Rutherford Lane
Stuarts Draft VA 24477

The Vegetable Work

First Edition 2015

ISBN-13 **978-1511630429**
ISBN-10 **1511630426**

Image Processing by Philip N. Wheeler

Table of Contents

Dedicated to Hans W. Nintzel,
American Alchemist
and
Founder of the
Restorers of Alchemical Manuscripts Society
(R.A.M.S.)

Disclaimer

Liability: The publisher does not warrant or assume any legal liability or responsibility for the accuracy, completeness, or usefulness of any information, apparatus, product, or process disclosed. The publisher makes no representation as to the accuracy or completeness of the contents of this book and specifically disclaims any implied warranty of merchantability or fitness for a particular purpose. No warranty may be created or extended by written sales materials or sales representatives. You should obtain professional consultation where appropriate. The publisher shall not be liable for any loss of profit or other commercial or personal damages, including but not limited to special, incidental, consequential, or other damages.

hollandus

OPERA VEGETABILE

(The Vegetable Work)

b y

JOHANNES ISAACI HOLLANDUS

Translated from the German by:

Leoné Muller

RAMS

1978

Introduction

Philip N. Wheeler

Johan Isaac Hollandus (late sixteenth Century) is said to have been the two Isaacs Hollandus, father and son, Dutch adepts, who wrote 'De Triplici Ordinari Exiliris et Lapidis Theoria', 'Mineralia Opera Sue de Lapide Philosophico' and other works on Alchemy. The details of their operations on metals may be the most explicit that have been given in writing, and may have been dismissed by some because of this very clarity. John Read, a Professor of Chemistry, in his 'Prelude to Chemistry, an Outline of Alchemy,' dismisses the writing of the Hollandus pair in a few words, possibly because their clarity of detail led him to suspect a ruse.

I have found this to be one of the best and most easily understood books on the vegetable work.

Please refer to the work *Alchemical Symbols* (The R.A.M.S. Library of Alchemy Volume 21) for help in interpreting the symbols.

Hans Nintzel selected this work for inclusion in the R.A.M.S. Library.

OPERA VEGETABILE

PREFACE

<u>T</u>O THE LOVERS OF LIGHT AND CHILDREN OF TRUTH

The most noble and very dear mouth of our beloved LORD Jesus CHRIST has not only itself, in its own voice, addressed inexpressibly gracious and above all desire kind words to the poor human beings; so that those very well-meant words, which issued forth from his living mouth - when, in person or the fullness of the essential Godhead incarnate, he walked on earth in his most holy visible form - and were diligently recorded for our comfort by the evangelists, especially John, can easily drive away the tears of a righteous man who reads or hears them, on account of the great graciousness of that LORD, undeserved in all eternity.

But this holy, holy, holy Light, which illumines all men who come into this world, has also from time—to-time, in all nations equipped various persons adorned with gratifying, delightful gifts of the Light as messengers witnessing his out flowing goodness, and has presented than publicly to the world to serve it. Neither has he left us High and

Low Germans out from this highly valuable gift. We will not speak about other German men, rich in Light, such as *Albertus Magnus, Tauler, Paracelsus, Basilius Valentinus, etc.*, but will now remember only the excellent philosopher *Johannus Isaacus Hollandus*. Even without remembering him, this man's name is without doubt known in all Europe on account of his extraordinary illumination and incomparable experience in matters of natural science and philosophy which he had through the gracious illumination of the Eternal Light.

This is sufficiently apparent in his writings, what there is left of them, and which are held in very high regard not only by our (own) modest philosophers but also by those of many other nations. That is why there is a great demand by many fanciers for his other writings, of which it is known that they were also committed to paper but which are withheld to this hour by envious persons. The demand is especially for his Vegetable Work, from which he quotes over 200 chapters in other tractates written by him and already communicated to the world. The same for his Animal Work, which is probably not much smaller, and others which I have not yet seen; neither am I informed where they might be. Should there be someone who possesses them, or knows about them, I would request him, in the name of the fanciers who would be grateful for it, to release their communication into their waiting hands, and in turn to be assured of the same readiness on their part. If all those wonderful writings were to come together into the hands of sincere investigators of spiritual and natural knowledge, much benefit would probably accrue to them thereby. For he (Hollandus) had a special noble gift from God for describing the great *Mysteria* in detail in very simple terms and frequent repetition of the important points.

In addition, he has such a friendly and affable style which pleases the reader so very much and makes him feel inclined to read his works often and to bear them in mind, all of which brings in greater returns for him. Therefore, remembering all this, I could not but be helpful in the publication of these writings; also because some time ago there came into my hands some writings which, to the best of my knowledge have never been published, such as the secret book entitled: *The Hand of the Philosophers,* also *Opus Urinae,* which two books have not been printed; also *Opus Saturni,* which has been added to the Basilii Triumpant-Chariot; likewise *Opus Minerale* which has been published in the Latin tongue; also *Opus Vegetabile*, of which I do not know if it is a part of the above mentioned Opus Vegetabile *(sic),* or a separate work.

These just mentioned Low German manuscripts were received a short time ago by the son-in-law of Justus a Balbians, and then reached me. Thus I have first translated the little *Vegetable Work* into High German, collated it with another written copy, and amended it far as possible That is the work I am hereby releasing for printing for the sake of the fanciers In regard to the Hand of the Philosophers and Opus Urinae, I have to wait them will more copies of them come to Light, so that they may be collated and corrected should deficiencies exist in them.

True, I also remember that what is contained in this little Vegetable Work is also found printed in part from the German Book entitled *Alchimia Vera,* or *Mons Philosophorum.* Bit I sensed that there in it much that was garbled and not a small part that was left out entirely. For this reason, the printing of the Opus Vegetable has been hastened. In so doing, I *am* fully confident that the fanciers of the secret divine gifts will realize that I am doing it for their special pleasure, to which end, that is, to prove to them that my mind is assiduously bent on serving, I have already done the little Tractate written by myself in German and Latin about the Sendivogian third beginning of the minerals as the Salt of the Sages.

To many, this has not been unwelcome, as I notice. Nevertheless, it has also aroused misunderstanding in some - this, again, being my opinion. In it, I tried to reveal, they said, that I would in oral dialogue reveal to anyone who contacted me, the most difficult points of the philosophical work which has for so many centuries renamed secret. And which an untold number of the must learning and cunning men had failed to understand and that, in my verbal discussions, would not hold back any of them as secret.

This would indeed be a great sacrilege, because from the beginning of this art God had instructed his philosophers that the extraction and the use of this Divine *SALIS ARMENIACI* should only be revealed to his servants predestined for it and by none but him alone, Him alone who knows the hearts, under pain of a miserable malediction to befall he who would break this seal. I leave it at that. But what has been done out of goodwill in my little book, is meant solely for the other righteous lovers of God and His Art, since it may be considered the obvious work of Divine Providence that God wrought us together by

His favorable Divine will. For He alone knows the minds which is impossible for us humans to know.

I must also inform them that I know that a short time ago some (men) have supposedly been found who claimed that they got to know me and obtained my confidence. These claimed to have received the revelation of these secrets from me, and that the same revelation is again available to any other wealthy fancier for a sum of money. In order to counter this kind of fraud, I hereby declare myself publicly and swear by the highest truth, which is the LORD JESUS himself, that I have not revealed to anyone in the world the whole foundation of the Matery *(sic)* for the Philosophical Work. If someone had boasted of some revelation by me, or would boast about it in the future, seeking to profit thereby monetarily, such a one would in truth be a wicked cheat and a liar. And may he suffer what the Apostle Peter said to those who think that God's gifts can be bought with money. Having said this much, let this matter also rest.

Now to return to our love — and praiseworthy Hollandus. Many a man may wonder at the mighty powers he ascribes now and then to the *QUINTESSENCE,* including the casting out of the devil from those possessed, and working such unspeakable miracles beyond all reason that it appears unbelievable to many. I cannot give these people any other testimony than to state that it is the pure, unadulterated truth. This is so not only of Hollandus, but also of the other philosophers who had actually possessed the <u>Quinta</u> <u>Essentia</u>. This has often been experienced and found to be true, as I shall show at the end of this tractate. All this can easily be believed by one who considers and heeds just a little, the super—abundant, inexhaustible love of God for His image: man. Yea, he will find that that is nothing in comparison with the depth of love - predestined to the earnest believer on Christ, even before the foundation for the creation of this whole world had been called forth and laid. About this too, I must be silent, since no eye has yet seen, nor ear heard, nor has it entered into anyone's heart, what GOD has prepared for those who love Him faithfully.

Therefore it is not possible to ask for anything better of God but that one should love Him in (the person of) the LORD JESUS CHRIST with all one's heart, all one's soul, all one's mind, and with all one's might, and likewise all other men as one's neighbor who is the wonderful dear image of (our) beloved God. Whoever possesses this great treasure (*i.e.*) that he finds some love for God within himself, let him give thanks most humbly for it; for without His Grace he would not have it. Let him also pray most earnestly for an increase of it (love) in himself and others. At the present time, when the aged body of the Christian Churches has almost grown completely cold, it is very necessary that some rouse themselves to fervent prayer and, together with men of weak faith, call day and night to our Father in Heaven with the following or similar humble sighs:

O Glorious, most blessed Creator, dearly beloved <u>Abba</u> Jehova Zebaoth. I, thy unworthy worm and poor mite, am not worthy to think of Thy almighty, terrible Name, much less call upon it with my impure lips, because Thou lettest Thy wise servant bear witness that before Thee, O Father, the whole world is like the needle of a balance, and like a drop of dew. What then should I, insignificant little worm, be conceited about, I who am not even a mote in a sunbeam in such a great world. And yet, Thou living fount of Love and kindness, hast bestowed so much mercy and Grace upon me, a mean creature from my mother's womb, that I can never count it all, let alone thank Thee for it.

I thank Thee nevertheless for such great, untold mercy, with all my heart, in deepest humility, especially for having bestowed the Grace upon me to love Thee dearly in my weakness. That love from Thee also impels me to do gladly what is dear to Thee, that is, Thy noble and solely good will and pleasure. I am a poor sinful man, however, and cannot do anything except Thou givest me both the willing and the accomplishment. Therefore, I beg Thee, I whine unto Thee, and implore Thee, by the dear name of JESUS CHRIST, Thy slaughtered lamb, to grant some mercifulness to me, Thy miserable creature, sanctifying me to (being) Thy faithful servant. Since Thou, knower of hearts, knowest - and I also swear to it before Thy omniscient Majesty and before Thy Angels - that I desire or demand no other blessedness than that Thou makest of me, Thy faithful servant, letting me do pleasing service for Thee and my fellow men, and helping to increase the honor of Thy glorified Name.

Do let me see with my eyes and let me hear with my ears that Thy Holy Name is hallowed over the whole world. And do away with that infamous Satan and the rapacious wolves and tyrants which only choke and dissipate Thy poor sheep, and thus desecrate the will of Thy mercy. LORD JESUS CHRIST, at the right of the power of God, who art with us all the time till the end of the world, these my prayers must be heard in virtue of Thy Almighty Name, as truly as

Thou hast said that there shall be one shepherd and one fold.

God of Abraham, Isaac, and Jacob, remember the promises Thou hast made to Thy servants, that Thou wilt soon bring them along by Thy Almighty Arm, and complete and establish the Kingdom of Thy Anointed. Of all this, may I obtain the simple little part of yet becoming and remaining Thy faithful servant. To that end, may Thy terrible Holy Name, Jehovah Zebaoth, help me, and the most dearly beloved name of our blessedness, JESUS CHRIST, who will eternally put to shame all Thy disobedient creatures. Amen. Amen. Amen.

In such a way, beloved brethren, let us humble ourselves under the mighty hand of God, because the prayer of the humble has always pleased Him, and, in His Most High seat in Heaven and on Earth, he has cast His gracious eyes solely upon the lowest. Let that be an unforgettable Memorial for us throughout life and a continual reminder to learn from the Lord JESUS CHRIST to be gentle and humble with all our heart. May he give us that attitude and let Him ever protect us under His wings.

Done in England, on the first of the year 1659.

From one who owes a service of love to the reader.

J.F.H.S

FILIUS SENDIVOGII

THE VEGETABLE WORK
JOHANNES ISAAC HOLLANDUS

CHAPTER I

Now I will teach and describe the secret of the arts, which secret is at the heart of all secrets hidden in the art of alchemy; since one will here understand the wonderful works that God has accomplished in all the things he has made out of the four elements. For I shall here teach you to know the spirits of herbs, trees, and all growing things; how to separate them from their bodies, and also how to purify the four elements and to restore them to their first being and their perfect power; that is, that when the elements are purified, how they can be put together again and make a perfect and fixed body of them, which is then glorified and has a miraculous effect. Enough of this for now.

CHAPTER II

And now I will teach you how to draw and make *Sal Ammoniac* from all growing things; for from all things that have the four elements in them you can extract *Sal Ammoniac,* because the spirit of all things is *Sal Ammoniac*. That is why *Sal Ammoniac* is designated and pictured as the sun. For the sun is the supreme sign and the most powerful planet of heaven, since the sun lights up everything in its essence; and it warms in the whole world, and does much more which is impossible to describe here. That is why *Ammoniac* is compared to the sun and is a wondrous thing, because without it nothing in alchemy can be brought to perfection.

Just as no thing or fruit can ripen without the sun, no work can be brought to perfection in alchemy without *Ammoniac*. For *Sal Ammoniac* can unite all things that are antagonistic and cannot be mixed, so that afterwards they mix and conjugate. That, therefore, is one of its capabilities, in accordance with the following example:

The sun has so much power in the mountains and in the minerals that, by the heat and power granted him by God, it can level and unite all unlike and antagonistic things, as big as they may be.

For what is more antagonistic than heat and cold, dryness and humidity, water and fire? Nevertheless, the sun is able, by his heat and power, to level all those things and to unite them in such a way that they will nevermore separate. That is *Sulphur* and *Mercurius*. Sulphur is hot and dry; Mercurius, however, cold and humid. But the sun, through his heat and power, unites Sulphur with Mercurius, from which then gold, silver, lead and copper are generated. According to where the minerals lie and the earth is good or bad, the metals are also generated. The sun, however, is he who must cook everything, and he cooks the metals better in one place than another, because he shines more in one spot than in another, since one area is more temperate than another. That is due to the planets under which a country is situated. In a place where the sun is too hot, he cannot cook moderately enough on account of the great heat, but finally the cooking will take place nevertheless. Instead, where the sun is too cold, the cooking is done slowly. But where the sun is temperate, that is neither too hot nor too cold and the place is under a good planet, by which is understood a well—tempered country (a country with a good climate), where there is good earth, the minerals are cooked moderately and generated into gold and silver.

It is the same with *Ammoniac*. It must unite and bind all antagonistic things, mix and level them by its temperate heat. For where the *Ammoniac* is too hot, it will cook all the longer; if it is too cold, it will cook proportionately longer; but if it has the right temperature, it will cook more gently and therefore accomplish a higher projection.

CHAPTER III

How is it to be understood that one kind of *Sal Ammoniac* is too hot and the other is too cold? You have to understand it in this way: Herbs are unequal; one is cold, the other hot. Nevertheless, they both have *Ammoniac* within them. Yet one kind of *Ammoniac* is better than another. Even so, both have power to cook in this art and to make a projection and a connection; that is, to mix and unite two antagonistic things. But it may also happen sometimes that the *Ammoniac* is not well-cleaned or purified, that the untempered heat that herbs have within them is not properly killed and purified. When then *Ammoniac* is made from them, the poisonous heat is detrimental to it, so that the projection becomes the smaller for it.

Likewise with cold herbs. Should anything remain in them, the projection becomes the more insignificant and small. That then is the difference: One kind of *Ammoniac* does not make the same sort of projection from one species or one kind of herb. Such, however, is the fault of the laboratory worker who has made the *Ammoniac*. (It means) that he has not purified it well enough and has not drawn it off often enough. For he who would make the *Ammoniac* correctly, must draw it off and off, often, till nothing remains and

24

it becomes as white as snow. Then the *Ammoniac is* of the right temperature; then the evil, poisonous heat which the herbs or species (other things) had within them will not be a hindrance when the *Ammoniac* is made; or if the evil, poisonous cold was in the herbs, it will no longer be a hindrance, because now everything is cleansed and of the right temperature, having left behind everything that was not of the right temperature.

CHAPTER IV

Further, my child must know that all things GOD has made from the four elements must die, be annihilated, and cleansed, but are afterwards again created and again born. Thereafter, the recreated and reborn things will never again die or be annihilated; nor will anyone be able to spoil or annihilate them; no fire can burn them. But they will henceforth last into eternity, because they have reverted to their prime power, given them by GOD when he created and made all things at the beginning.

As a consequence of the sin committed by our first parents, Adam and Hevae, GOD the LORD was so angered that he spoiled the four elements and made them corruptible, so that they must now all die and come to naught. Then GOD the LORD himself spoke to Adam as he was giving him this command, saying: "Adam, of all the things that are in paradise, you may eat, except from this tree, because it is the wood of life. And I am telling you thus, Adam. If you eat of this tree, you and everything created of the four elements shall die, that I swear!"

GOD made an oath there of which we have all become well aware. That is why everything created and made of the four elements must now die, including human beings, all animals, and everything that has received a body or has sensitivity in it; yes, also everything that has no sensitivity in it, such as herbs and trees and everything made of the four elements. For if some may not have any feeling, they yet live and bring forth their fruit. Just like the sensitive creatures, herbs bring forth fruit and their seed, or whatever nature GOD has implanted into them. Trees carry apples and pears, or that which GOD the LORD has bid them do. Thus live all herbs, all trees and all other things created by GOD the LORD from the four elements; and God the Lord has given a spirit to all trees, all herbs and all other things created by him out of the four elements. As many a thing or herb there may be which he made of the four elements, GOD the LORD has given to each of them its special spirit of a particular power and a particular nature. Each has its particular body and shape, and that same body is simultaneously made of the four elements. One herb is cold, the other hot; the third is humid, the fourth dry, according to whether each has in it much or little of one or another element. That is how it happens that one herb is cold, the other hot.

But spirit is something else which GOD the LORD has given to all things. And as many different things GOD the LORD has created contrary or antagonistic to each other, as many different spirits he has also infused into them. Each spirit has a miraculous power and virtue for a special sickness; and each spirit has the power to accomplish some special work, with the help of other *species,* as well in metals as in human beings, for GOD has created all things in behalf of man.

CHAPTER V

Accordingly, each spirit performs a special cure in man or in *Mercurius.* The spirits, however, have not much power because of the impure body by which they are sullied and surrounded. For it (the body) is made of the four elements, and the four elements are so impure and dirty that the spirit is quite unable to apply the same power and do the same work for which it was created by GOD. In addition, the time

of the world is up, and it is now becoming too old
and weak. The sun and the elements are losing their
power, and the elements are becoming so infected
(polluted) and impure that the spirits, on account
of the impurity of the elements, can only have an
insignificant curative effect. Of course, at the
time GOD the LORD pushed and drove Adam out of
paradise, herbs had greater power than they have
nowadays. That causes the world to grow old, and
therefore people do not live as long as in former
times. That is due to the present old age of the
world and the fact that the sun and the planets are
losing their powers. Because of the age of the world
and the pollution of the elements, the spirits in
the herbs are so overcome by the pollution of the
elements that they can no longer generate their
powers in man as they did ages ago. If now they
would still manifest their powers as they did long
ago, human beings would still live today into their
200 years and beyond, and they would in everything
have the same powers as they had previously;
although people are now also weak and delicate and
could not tolerate the spirits of the herbs if they
still had the powers of years ago. They would
certainly have to take and use them tempered. If
they did that, they would live even much longer and
stay younger.

That is why all herbs and other things have to be killed, annihilated and reduced to powder and ashes and finally to water. Afterwards, the soul, or spirit, has to be infused back into them and a perfect body must be made of them. Then you have an earthly treasure that is better than gold and silver and precious stones. For you have a perfect glorified *corpus* (body) which will never pass away but will last eternally, passing through all things. And where it passes through, it will not leave any corruption or disease at all, but it will heal that through which it penetrates before leaving it. Not only will it make it healthy but much healthier than it had ever been, and it will also keep it healthy from then on. Yes, if it had never been healthy before, it would be made healthy thereby and preserved. That is why I may well say that it is above all earthly treasures, and you will notice yourselves what a treasure you have here. To those who understand, enough is said hereby. Enough of this.

CHAPTER VI

In addition, my child, you must know: The fact that I said in the previous chapter that one must kill and let the herbs die, and make a powder or ashes of them, is to be understood as follows. One must draw off the evil, impure humidity, or let the herbs dry of themselves, which is best. Or else draw it off in Mary's Bath. The evil moisture is that which hinders and overcomes the spirit most, so that it cannot generate its power in man or in metals that is, in MERCURIUS. The miserable fools work in their laboratories with the humidity; and although there is no danger that anything good be done or found by them, the evil humidity robs them of that which they are seeking. They put the herbs to putrefy, then draw them off and operate with the putrefied matter they have drawn off. They work hard, but at the end everything the poor wretches have done, is lost. They cannot fix it. That is due to the putrefaction and the evil moisture with which they worked. Then everything the poor fellows have messed about with is lost. Thus, the art seems impossible to them. They start despising and slandering it.

Why did I say this? So that you should or may know what causes their mistake and what they are lacking. You might think that they separated the four elements and (wonder) what is the reason that the spirits do not have their powers, since, after all, the elements were separated and purified. But that is done by the evil wateriness with which the herbs are putrefied and with which they have worked. That is the reason why. Understanding it, beware of the evil wateriness. An accident known beforehand is easier to prevent. And that is why I have told you so.

CHAPTER VII

Further, when you have drawn off the evil moisture and the herbs are dry, draw off the spirit *per descensum,* as I will teach you later on. After that, calcinate the *corpus* as white as snow. Then you have two natures, that is, body and spirit.

Now dissolve the Spirit in *Aqua vitae,* which is very good, or in good distilled wine—vinegar that is pure. That is the liquor with which you should work. In the same way, dissolve also the *corpus,* as I instructed you concerning the spirit. When coagulated, you have two kinds of *Materia*, to work with each in a special way. But you still have no perfect glorified *corpus,* since the spirit is not yet united to it, and they are not yet married together and connected (or joined). That is also why they have not yet got their perfect power, although they are clean and pure. And although they have already been ashes and water and have now come alive again, they still differ in as much as each is alone by itself. Nevertheless, each has great power by itself, which they show each in its own way.

CHAPTER VIII

In order to understand better what is being said, you must understand that there are two kinds of sicknesses in man and also in MERCURIUS. In man, there is a disease called "spiritual sickness", and still another called "befalling sickness." I do not, however, mean infirmities of the soul when I speak of "spiritual sickness." The meaning is that people may well have different dispositions in their bodies. But what I call "spiritual sick-days" are those which befall people by accident or chance, such as, because of anger, fantasies, regrets or grief, or unexpected mishaps, or because of losses, or from much studying, or if man gives too much to do to his senses, or from anxiety or troubles of the senses, or from fear and fright, which befall people or are caused by them, and from many more other things which it would take too long to write about here. All of them are spiritual sicknesses from which serious diseases can develop for people, such as, bad fevers, bad hot sicknesses and others too long to relate here. All such infirmities must be treated with the spirit of herbs. When they are thus prepared on their own, the spirit has the power to purge them.

CHAPTER IX

Further, there are in man other sicknesses which are called befalling ones. They come from much eating and drinking, or from bad food or drink that people take, or from overstuffing themselves after suffering great hunger and thirst. From that, bad sicknesses arise and *Apostemen,* bad water swellings on the liver and the spleen. The lung is spoiled by harmful cold or heat. In addition, the kidneys are made sick by excessive eating; and by too much lewdness, whereby the blood of the loins is lost. These and similar sicknesses, of which there is a great number, are called befalling sicknesses. They are cured with the body of herbs after they are prepared in this way, and that is their power and effect.

But when the body and the spirit have been put together and unified, which means, brought together to *fixity,* they have such a miraculous effect that one cannot describe it with any pen what power and virtue they have. There is no Master in the world who could fully fathom the might and capabilities GOD the LORD has granted them. Thus, the spirits of all herbs, trees and *species* are so noble that all doctors in the world could not understand the nobility of a single spirit, even if it were from the meanest little herb GOD has created in the world. How then could they know the powers of all herbs, trees and *species,* since each *species* has a special nature and spirit, and one spirit is always nobler than the other.

CHAPTER X

I am well aware, however, that among all insensitive spirits none is noble and vigorous as the excellent spirit the noble vine has for Almighty God has foreseen from eternity that the noble Wine should grow on it, which was to be transformed into God's blood and body. Thus the sages know well that the spirit of wine is above all spirits of the other herbs. That is why the ancient philosophers did not find better powers in any herbs, trees or *species* than in the spirit they drew from the wine. That is why I may well say that the noble spirit of the vine is the noblest and best among all things. Therefore, the spirits are one nobler than the other, but their utmost powers cannot be gauged or found by anyone but GOD alone. The fattest herbs, however, which carry seed, are best to make *Sal Ammoniac* with. After these, the hottest herbs (are best) from which the most vigorous and strongest *Sal Ammoniac* is made.

CHAPTER XI

If then you wish to prepare a medicine which is to affect metals or *Mercurius,* you must take the hot and vigorous herbs or roots, that are of no use to either human beings or cattle, and prepare an *Ammoniac* from them, and an *elixir,* as I shall teach you later. If, however, you wish to prepare a medicine for men, take good, lovely herbs that people can use. Prepare a medicine or elixir from them, with which you may work on people. Then you will effect such miraculous cures in people that the whole world will wonder at you, and everybody will wish to see you. Enough of this. Understand me well, however, concerning the *Materia* I have hinted at from all sides.

CHAPTER XII

My child must further understand that I have said and taught in previous chapters how one can recognize the nature of herbs, separate their spirits from their bodies, and what one is to do with them. That I will now explain and teach better.

Know, therefore, that there exists still another spirit or *Ammoniac,* that is, coming from *salinic* things which is also *Sal Ammoniac.* For the spirits of all insensitive things, when they are separated from their bodies, are called *Sal Ammoniac*. Know therefore that the spirit of all *Salien* (salty matters) is called *Sal Ammoniac,* but it is not the *Ammoniac* meant by the philosophers. Of that (the one meant by the philosophers) they prepare elixirs, but one cannot make elixir from the other. It is the philosophers' soap and washing water, with which they purify and cleanse the bodies. And with it they dry the elements of their evil moisture. In addition, they dissolve the bodies with it and conjugate things with it which are contrary or antagonistic to each other. It is a volatile spirit, one that goes in and out; and if it were not so the elixir would not come in.

In this *Ammoniac* there are also many things which cannot be described, since one can do miraculous things with it if it has first been fixed. But that is not necessary for this work. But anything the *Ammoniac* from the salts can do, the *Ammoniac* drawn from the herbs can do also; and an elixir can be prepared from the *Ammoniac* drawn from the herbs without adding any other *species,* which cannot be done with the *Ammoniac* from the *Salium*. But one can well prepare such an *Ammoniac* with other *species,* so that *Mercuris* can be dissolved with it in water; the same for all other metals and things, provided one proceeds as I have taught elsewhere. Enough of it for now.

CHAPTER XIII

Now I will further teach my child and describe the powers and virtues that the herbs have when the elements are purified, cleansed, separated, calcinated, made to water and afterwards again put together and fixed, and a glorified *Corpus* has been made of them. Neither I nor all the doctors of the world can sufficiently comprehend the powers and nature which I, and the journeymen with whom I work, have seen and tested, and of which other Masters have told me. For only GOD alone can comprehend the extreme powers that herbs possess when they have thus been prepared and made into a perfect *Corpus.* Nobody can know it but GOD alone. It is known to him and to no one else.

CHAPTER XIV

In addition, you must know that there are many mistaken men among those who work in this art, whose error I mentioned briefly before, when I gave information on the bad wateriness the herbs have within them. With that they work and let the herbs putrefy with it. Afterwards, when they have distilled that wateriness from the herbs, they call the Element of the water. But through it they cannot reach perfection at the end, of which I shall speak more later.

Here I shall teach how to separate the elements. You must know that there are many kinds of separation of the elements, since there have been many artists who all thought they knew the way of separation. But one (way of) separation is better than the other. Yet both are good. Among the learned and the unlearned one finds foolish men who also wish to perform these works. They begin to work in laboratories and imagine that they also understand the art of separation of the four elements. Then they say that they have separated the elements from each other, and each in a special way. They imagine that they have performed great miracles, saying "We have made the quintessence." True, they all drive out many sicknesses from people. That is certain, because it (the quintessence) has a great power and virtue in it, more than they know.

But the miserable fools fancy that they have made the quintessence and separated the elements one from the other, is nothing and sheer deceit. True, they have a great medicine, more so than they know themselves. But that they pretend and say that they have made the quintessence is far from the truth. You poor fools, you have no quintessence. The quintessence is quite another thing than you think. It is your glorified *corpus* brought to perfection and fixed, and lasts throughout eternity. Whoever has such a thing can say that he has the quintessence. He has an earthly treasure that is better than a kingdom. It is a gift of GOD, which he bestows especially on his friends. Happy the man who acquires it and knows how to use it well for the blessedness of his soul and the benefit of the poor. He will fare well in this world and the next.

Instead, those who use these gifts of GOD differently shall have their troubles here in this world, and later suffer infinite tortures in the eternal hellish torment. Take good care, therefore, that you use the art for the honor of GOD and the salvation of your soul. For I swear by the living GOD, who has created heaven and earth, that, if you use this gift of GOD otherwise, you shall not live long, and you shall be tortured in this world with temporary, but afterwards with eternal torments. Therefore, take care yourselves what you do. It would be better for you not to be born than that you had the art and misused it. Therefore, watch well! Enough said to those who understand.

CHAPTER XV

Now we will return to the separation of the four elements. About that there are many different teachings and arguments, to quote which would take too long. I will only conclude (the matter) in brief words.

First, all herbs have within them the four elements. Three elements are visible and tangible as water, earth and fire, but air is invisible. It is hard to separate the water from the air; yet it is possible to do it by drawing it off slowly in Mary's Bath, on a small and gentle fire, so that nothing rises except the wateriness. If one were to give a strong fire, however, so that the water would boil, the air would also rise, a little or much. But the water can be separated from the air by a gentle fire.

The earth and the air are also difficult to separate from each other; but by a big, hot, strong and long lasting fire, the earth can well be separated from the air. Should one give a weak fire, the air would stay with the earth which would then join the element fire; for the element fire is the last one to separate from the earth. It must be separated from the earth by strong heat and a long lasting fire. For if something of the element fire were to remain in the earth, the air would not be separated from the earth; because fire and air cannot be separated, although many fools work in this art who are of the opinion that they separate the elements into four parts. They are mistaken. True, they separate four things, and then they believe that each is a separate element in itself. Oh, no, you fools! And although you are learned in writing, you are nevertheless more foolish than the unlearned. For if the latter have already gone astray by thinking that they have divided the four elements each by itself, you, considering that you are scholars, should by rights not remain in error, since you have sufficiently studied the books of the Masters of natural science as also other writings. Accordingly, I am surprised that you talk yourselves into believing that you can separate the elements one from the other, each by itself.

I do not speak of the true, trustworthy and experienced artists who understand the hand of philosophy, but I am speaking of some learned men, both clerics and laymen, who wish to work in this art and do not have the old hand of the philosophers, and are not familiar with, nor have been sworn into this art, will lose everything they employ in it. Never will some of them reach perfection, unless GOD would enlighten them miraculously. The devil has no power at all to teach anyone, as has often been experienced. Therefore, I know well that he has no power. That is why it must come solely from GOD, and the art is therefore called a gift of GOD. Happy the man who has it and uses it rightly!

CHAPTER XVI

Accordingly, I call such people, whether they be clerics or laymen, great fools for imagining that they have understood the art with the help of some books which deal with the art in parables. Thereupon they proceed to work and lose all their expenditures, because of which they often land in poverty and despair.

Even so, they do not stop. I have seen this myself in clerics and laymen who used up all their belongings, becoming poor because of it, so that they may rightly be called fools.

Of course, the unlearned cannot be blamed, for they know not what they do. Afterwards, when they have become poor, the art seems to be an impossibility to them, and according to their belief, it is also true. For it is impossible for idiots and the unlearned to perform the art. How should the art be performed by an unlearned person? Such a man could not understand it. That is why an unlearned person believes that it is impossible to perform the art, and in that they believe the truth, since it is true that the art is impossible for such people. Why have I said all this about the learned and the unlearned? I am doing it because of the separation of the four elements, since they say that they are able to divide the elements one from the other, each in a special way, so that one element is not mixed with another. To do that is impossible for them. They must be mixed, air and fire. But that must be done by the right Masters who have had the hand of the philosophers and understand this. They may well be able to divide the four elements, each to its own; but no one else in the world (can do it), unless it were the will of GOD, as I said before.

CHAPTER XVII

It is here not required that the elements be divided from each other, each alone in a special way, but only to cleanse and purify them. To that end you must separate the water from the air, the earth and the fire. Then you must separate the air and the fire from the earth, and purify the air. Fire in itself is pure, but fire must be worked upon together with the air, by means of the air that is in the fire. Then purify the air by calcination, as I have taught before and shall teach still better later on. After you have drawn from the herbs the evil wateriness, keep them standing closed in the fire. Give them a small and gentle fire for 12 hours, somewhat stronger every hour. Then there will go from them a white, red or yellow smoke, according to the spirit of the herbs. For there are some herbs that have a red spirit, but all ordinary herbs of the world have a white spirit. That a few have colored spirits would here take too long to relate, nor would it serve this work.

With this gentle fire one gives it, increasing it gradually for twelve hours, the element air will in the meantime go over. That is the white, or the colored, spirit. The old philosophers call the element of the air *Ammoniac* which word *Ammoniac* has much in it. Therefore, they call all spirit *Ammoniac*. This work must now show the reasons therefore.

When now the air, or the white spirit, has been drawn over cleanly in such a way, you must heat stronger for another 12 hours, increasing every four hours; still stronger for 20 hours; and as strong as you can during the last four hours, so that the barrel stands in the heat. The oil will go over within that time, mixed with the air, or whatever it is to be called. Then you have to draw the three elements from the earth. First, the water, then the air, and following that, the oil or fire. Now you must calcinate the earth in an even heat for three days and three nights. Give a strong fire, as hot and strong as you can. Then take it out, and you will have the earth pure and clean. After this also purify the air and the fire, and give it its water, pure and clean. That is the vinegar or brandy ("burnt wine"), well cleansed and purified by distillation, as I taught you previously.

Now, put the four elements together and make of them a perfect *Corpus*. Now the elements are again gathered and united with each other, joined and fixed. Now it is a perfect glorified body which lasts imperishably into eternity. If now all artists who were ever born, or may yet be born, would come together, they could never again separate the elements from each other. Yes, all the devils in hell now have no power to do that, neither anyone else but GOD alone. Only now may you say that you have the quintessence, which is indeed a gift of GOD.

CHAPTER XVIII

Now we shall see from what the pure glorified CORPUS
is made which is called the quintessence. It should
be prepared from herbs. You ask from what herbs? I
say from all the herbs that are in the world, hot or
cold, dry or humid, as they are, and even if they
are poisonous because of great heat or cold. Some
herbs, however, require more work than others;
because the good, natural herbs that people are used
to eating, need not be sublimated, distilled and
calcinated as much as the strong, hot and dry herbs
or those that are so very cold, moist and poisonous,
and are totally inedible for people. The evil cold
or heat has to be removed from them by sublimation,
distillation, calcination, dissolution and fixation,
and by calcination, dissolution and coagulation.
That has to be done so often that the bad poison the
herbs have within them disappears and a great
medicine becomes of them; yes, a great elixir. For
the stronger and more poisonous the herbs are, the
higher projection they make, provided they are
brought to perfection through the hard work of the
Master who has treated them the way I taught before.

One should cook a snake or dragon into a basilisk
through sublimation, distillation, calcination,
dissolution and fixation. Or, as one should say of

such a work, the venomous animals are finally to turn into a great medicine and the elixir for metals and human beings. One can in that way kill the venom in venomous animals and turn it into a great medicine and elixir. What is most suitable to do that are the strong herbs which people cannot use. In this way they can be brought to perfection. Likewise the herbs that are naturally good.

CHAPTER XIX

If you should now ask what and how many herbs you should take, hot or dry ones, or of what nature they should be, I answer by telling you that you should take many kinds of herbs, as I said before, hot and cold ones, dry and moist ones. The more different herbs are used together in the work, the better it is and the more power and might the quintessence will have. When they have been brought to perfection, that is, to their highest power - for GOD has created nothing without a reason and has bestowed on each thing a special power and virtuous nature, as I have already indicated; therefore, the more different herbs you take, the better it is, since the more people are together the more courageous they prove to be, and one leads the other; likewise with the different herbs — when they have reached perfection and then get inside man, they produce miracles, because each herb does its own, and where ever they get into, they do not leave anything imperfect in man so that no kind of disease remains in him. For if any infirmity which had befallen him in his lifetime and he was not born with it, would remain in him, the medicine could not go by the name of quintessence. Even if anyone should be possessed of the devil and he were given the quintessence, the devil would be forced to leave the man immediately, because the evil spirit in particular cannot stand the quintessence. These are the reasons. The devil is the real cause that the elements have been made corruptible by GOD the LORD and that they have been altered. That is why he cannot suffer that the elements should again reach their perfection as they were before our first parents Adam and Eve were brought into sin by him, on account of which the elements have afterwards been corrupted with impurity and decaying.

That is why the enemy cannot stay where the quintessence gets into. And whoever carries the quintessence with him, is protected from the devil. There are many reasons why the devil must flee from the quintessence; but it would take too long to describe them here, because there are many such reasons. Briefly, there remains no evil where the quintessence gets into, be it from heat, cold, dryness or humidity. The quintessence corrects everything, because the herbs which had previously been hot and dry, and with them cold and liquid people had been helped, are now reversed; the outside (has been turned) inwards; the inside, outside. They died and have risen again and have become alive. Henceforth, they will never again die. The heat and dryness which were their nature before, have now been reversed, so that now hot, gaunt people can be healed with the same herbs which were previously poison for them.

It has now been changed into a medicine. The cold, humid, although it had already been curative before, is now an all the better, and curative medicine. As is the case now with the cold and moist herbs with which one used to cure hot, dry sicknesses, it is now in reverse with the hot, dry herbs.

CHAPTER XX

And further, if a man were to take everyday a little of the quintessence in wine, with his food and drink, or in the morning, he would not die, unless nature would die of its own. He would remain in the same beingness and condition as he was when he began taking the quintessence, and his face would not get older nor his members more awkward, stiff or bent, because the quintessence would drive out right in the beginning the evil which man might have within him. For wherever the quintessence gets, no infirmity or evil can remain. That is why it is called quintessence or elixir. As soon as it has consumed or driven away the sickness, it makes the blood youthful again. When then the blood has been rejuvenated, all members again become well, quick and strong and remain always so. Neither need he be afraid of any kind of poison, for no poison can harm such a man.

I, myself, have seen that an ounce of arsenic was put in a glass of wine vinegar and given to drink to a man who had forfeited his life. Thereupon he was given a glass of wine in which there was as much of the quintessence as a grain of wheat is heavy, and the poison did not harm him. He told me that he knew nothing of it, that he had felt nothing, and one month had already passed when I asked him. That is why such a man is safe from Poison and from the fire of the plague, no matter of what kind it may be, seeing that the fire of the plague is of different kinds. I have, myself, given it with my hands to more than a thousand persons who had the fire of the plague. They were lying there and raving like madmen. As soon as the quintessence had passed through the throat, they became immediately healthy, and the fire ran off them as black as pitch and was stinking so much that nobody could tarry there. The same people told me that, as the black matter was discharged from them they did not feel other than if their behind had been burning with great heat, or if one had driven a hot iron into it.

Likewise, I gave it to twelve lepers. They were so leprous that they could not be recognized at all. Within nine days they became healthy and good looking like a newborn child, although one could still see the scars where the lepra had been bad. Within a month those had also disappeared by taking as much of the quintessence as was equivalent to the weight of a grain of wheat.

Also, I have given it to about one hundred persons on their deathbed. They had already been given up by the physicians who said that they were to die and that it was impossible for them to live one more day. I gave them the quintessence and brought them back to health within 24 hours. Thereupon, I gave them a *Confortative,* and they went outside again within 8 days. They said they had not been as healthy during all of their lifetime, and they thought they were flying when they were walking. For the quintessence had driven off everything bad in their bodies, and the tonic had made them new blood and given them so much strength that they felt so well that they did not know where they were.

Also, I have helped many a pilgrim who had come down with S. John's, S. Cornell's, S. Hubert's and other troubles. I have also driven the devil from possessed persons, because the devil must leave where the quintessence reaches. And aside from those, I have accomplished more than a thousand deeds with the quintessence, which it would take too long to relate here. But I will close with a few brief words and say that where natural death is not present, or the hand of the Almighty, meaning that GOD wants to trouble man on account of his sins, no infirmity can enter man's body that could not be removed by the quintessence, and that within nine days. But enough of this.

CHAPTER XXI

Now we shall see how one should make the quintessence, or a glorified CORPUS, and what are the characteristics of a transfigured body; in what manner one is to proceed, and what belongs to such a Corpus. A glorified body must be above all (free from) infirmities, perfect in all parts, clean and pure. It must penetrate all things; nothing can resist it, because spirit and body which before were against each other and strangers, are now great friends. They are married and joined together, so that they will never again separate from each other, since Spiritus and Corpus are no longer two but one single substance and an inseparable thing. As long, however, as it is more than one thing, of which the elements can be divided from one another, they are not fixed but contradictory to each other. But when contradictory things are prepared and brought to their own nature, and are throughout reversed, as I have taught in *Mercuris,* and are subsequently together so that one mixes with the other through the art of the laboratory worker, only then are they made inseparable, so that one single thing remains. Although many things have been brought together and made inseparable, they are yet afterwards no longer many things but one thing. Then it is fixed, incombustible and invulnerable.

Or do you think that when GOD will raise the bodies from the dead, create them again and restore to each his body, everything will be as it was before? No, but the body will be created again and made new, completely reversed, invulnerable and immune to sickness. Those who now have fat abdomens and bad, watery stomachs or bodies, do you think that GOD the LORD will give them again such obese abdomens or unhealthy livers? NO! Not at all. GOD will restore them to their first nature and their first beingness. He will take from them everything unclean and give them what is required. He will remove every bad wateriness and dry the bodies and then moisten them with heavenly dew; that is the Heavenly Humidity with which GOD will moisten the bodies. He will take away all combustible fattiness and dry the marrow of all the bones and members without obesity. And GOD will fill the tubes of the poor, the thighs and the skull, with fat spirit. That is the fattiness which the transfigured bodies receive after the soul again enters the pure body. Then it is one (thing) and they will never again part from each other. It is one thing. The body is spirit; and the spirit, body. Then it is fixed and one quintessence.

To what purpose have I said all this? So that you should understand the work of the herbs all the better. For just as I have said that GOD the LORD will recreate the bodies and remove from them their bad wateriness and water them with Heavenly Dew, just as he will detach from them their combustible fattiness, making them fat again with the fat spirit of the souls, thus you must also do with the herbs. You must take from them the bad wateriness, as I have taught in the beginning chapters, and give them again the burnt vinegar or wine, which must be good and straight, as I have previously instructed you. That is the Heavenly Dew which I have in mind.

CHAPTER XXII

When the body has been separated from the spirit, or the elements from the earth, or as you wish to understand it, you must skim off with a feather the oil that swims on top, and that oil must not again come together with the earth or the Corpus. Nothing else must get together with the earth than the spirit which has been drawn off from it, or the *Ammoniac,* as you understand it. It is the fat spirit that I mean, which GOD the LORD will give back to the bodies and infuse into them. Thus you should understand, when I speak, that you must take from it what is too much in it; and you must restore to it that which it is lacking. By that I mean that you should take from it the bad (thing), which is the bad heat or poisonous cold, through distillation, or through sublimation, calcination, dissolution and coagulation. Repeat that so often till the badness which the Materia contained is driven out of it and it becomes natural and pleasant. In that way you should understand what I mean when I speak of a thing which you should rid of what is too much of and give what it is lacking. That is my opinion and that of all philosophers. When they refer to it, they wish to have it meant and understood the way I have instructed you sufficiently.

CHAPTER XXIII

Now we shall return to our first work, that is, how to join *Corpus* and *Spiritus*. When, therefore, you have parted the spirit from the body, you must remove the oil with a feather, so that none remains on it. You must put the oil in a glass, well-sealed with Luto Sapientiae, and preserve it well, because you will perform miracles with it when it is prepared the way I am going to teach you to prepare it. The unprepared oil, however, is too unnatural and poisonous, for in it there is still the poisonous heat which the herbs have had in them. They must be driven out; then you will afterwards accomplish miraculous cures with it.

After you have removed the oil, put the spirit into an alembic; put a head and a receiver on it. Distill it with a temperate fire, that is, not too hot and not too cold. When you have distilled it, add the feces that have stayed in the alembic to the earth in order to calcinate them together. Put the spirit back into the alembic, put the head on it and the receptacle. Distill as you did at first. Thus, you must distill over and over ten times. Then the spirit, which was before poisonous, hard, evil, sharp and useless to anyone, becomes pleasingly sweet and natural. Now, however, it is pleasing, good and natural, so that its virtue cannot be described or explained. That is the fat spirit I mean when the pure, dry CORPUS is to be fattened.

CHAPTER XXIV

And now we will also give information on the earth, or the body. It must be prepared artificially, for when the element earth has been prepared and brought to its first nature and its own being, it does miraculous things whose power is indescribable; previously, the earth was black and useless, and all elements wanted to flee and be separated from it, because they were clothed over and covered by the earthly blackness, so that the elements could not use their nature. Now, however, after the earth has been brought to its highest power, or to its own being, that is when it reaches the state in which it was when it was created by GOD before Adam and Eve and he afterwards cursed the four elements so that they became corruptible and adulterated; when then you have brought the earth into its state of purity, the elements which previously fled from it now desire to be with it, as I have taught before.

Therefore, when you have drawn all the elements from the earth, put the earth into a long earthenware vessel, baked out of potter's earth, so that it gets heated through all the better. Put it into the furnace of calcination, and calcinate it for three days and three nights in as much heat and strong fire as you can give. When the three days are over, take it out of the vessel and put it on a stone. Rub it firmly with brandy ("burnt wine") out of the vessel. After that, put it into a glass barrel, pour more of the wine upon it, and place the vessel into the Bath till the earth is dissolved. But cork your glass well to prevent the spirit of the brandy from flying out; since the spirit is altogether too agile and too subtle, it would fly away invisibly.

When it is dissolved, let it grow cold and let it stand for three days to settle. Then take another glass, skim the clear above from the *fecibus* in it. Pour more brandy upon the feces and put it again into the Bath for one day and night, in order to draw the earth well out of the *fecibus*. Add it to the other clear wine. Do this three or four times till you have the CORPUS of the earth completely out of the *fecibus*.

After that, put the CORPUS into an alembic with a head, and distill the brandy. Then you have a very clean, white CORPUS. Thereafter, you can infuse the spirit into the CORPUS, and dissolve the CORPUS in the SPIRITUS. Then put the thus dissolved CORPUS into a glass together with the spirit, and put it to *putrefaction,* or into Mary's Bath for fifteen days.

After that, take it out; put the head on, distill, and test (to see if) whether something more is distilled than the wateriness. If you find that spirits are distilled, distill to the other side. Pour it again upon the earth, close your glass, and put it again to *putrefaction* or into Mary's Bath, for three days. Then put the head on the vessel, distill and test if something else than the wateriness is distilled. If something of the spirit is distilled, distill everything the other way completely.

Afterwards, pour it again upon the earth and put it to putrefaction for three days. You must do this so long till nothing is distilled but the wateriness. When nothing is distilled but the wateriness, it is fixed.

Then distill the wateriness off, pour it on again, and distill it again. Pour it on again. Do the pouring on and distilling off so often till it has absorbed all the water and is coagulated hard. Then rejoice. You have a glorified CORPUS which is precisely the quintessence.

CHAPTER XXV

Concerning the earth of which I said previously that you must again draw off the brandy that is good. Draw it off and do as I have taught you here. But it would be better if you were to pour the brandy back upon the earth and distill it off again; then pour it on again, as I have taught until now. And this so often till the whole quantity of the distilled wine were again infused into the earth, and you would again dissolve it, that is the CORPUS, in a good other brandy and put it again into the Bath, as you did at first; and again distill the brandy off, and pour it on again till it had sucked in everything; and dissolve it again as at first. And if you did this the fourth time, the earth would become so strong that it could not be described; neither could its virtues be expressed. When the earth is thus dissolved in its own spirit, as I have taught before, and becomes fixed with it, it does then ten thousand things where before it did one. This glorified CORPUS, or quintessence, prepared in this way, no King could pay with all his wealth for one pound of it, so great is its strength and virtue. Blessed is he who has it, and things will go well for him who uses it well. But he who misuses it will be tormented by GOD temporarily in this life and eternally in the next.

CHAPTER XXVI

Now we shall teach and show you another way which is better than the one I taught you before. Nevertheless, the first way is short and good. But this one is surer and better but requires more time and work. Yet both are good, and I have worked in both several times with my own hands, through the grace of GOD, as follows:

First, you must take, with divine help, any kind of herbs you wish, the more the better. Let them dry in a room shut off from sun and air, as I taught you previously in regard to the other work. When the herbs are dry, put them into a warm oven, not too hot, so that the spirits do not volatilize. When the herbs are so dry that they can be rubbed into powder by your hands, remove them from the oven and pound them in a mortar into subtle powder. After that, rub them on a stone with distilled wine, as small as you can.

Have at hand a large glass or earthenware vessel. Put the powdered herbs into it, so that they lie by one-quarter under the burnt wine. Stop the vessel with a cork. Then take one part wax, one part pitch, and one part resin. Melt them together in a pan to which you must also add one part of *Certissae* or *Miny,* and stir everything together.

Thereafter, take a strong hempen cloth, put it around the mouth of the vessel and tie it outside with a hempen string. Smear the string over with the *Materia* that has remained in the pan into which you had dipped your cloth. Now, put on it one finger's breadth of sand or pounded brick. In that way, the cloth cannot come undone in the Bath over the sand or the stone-powder, but lute it well. *Lutum* of a hand's breadth all around, and put a strong hempen cloth over the *Lutum*. Wind strings around it and then let it dry very hard.

When it is dry, pound ashes with the white of eggs and coat with it the cloth tied around the Lute. Do it also one finger's breadth, and let it dry in the cold. Then the vessel is prepared for being put to putrefaction, or in Mary's Bath. My advice is to put it into the Bath. Let it stand therein for 36 or 40 days. Keep the water day and night at such a temperature that you cannot keep your hand in it.

At the end of 40 days, let it grow cold during four days. Then open it, and keep ready an alembic which you can put into sand. Put your matter therein. Distill it, first by a very small and later by a stronger fire. Distill everything you can. Remove the *feces* that remain in the alembic and rub them on a stone with good, fresh burnt wine, so that they become impalpable.

Then return the feces to the aforementioned vessel and pour upon it what you have distilled. Close the vessel again as before, put it again into the Bath for nine or ten days; give it fire or heat as before. After that, let it grow cold. Then open the vessel, put it into the alembic, a head on it, and distill, as I have already taught you before.

Remove the feces, rub them on a stone as before, and return the matter into the alembic. Put the head on, and distill. Do that three or four or several times; as often as you do it, meaning, drawing off and again pouring on, as I am here teaching, its power grows and increases tenfold, which tenfold is each time increased tenfold. Therefore, do not grudge any pains. You will be rewarded a hundredfold for it. One should really draw it up and off (to multiply it) so often that it would at last achieve a projection as great as the Great Elixir.

FINIS PREPARATIONIS QUINTAE ESSENTIAE DUOBUS MODI.

(END OF THE PREPARATION OF THE QUINTESSENCE BY TWO
DIFFERENT METHODS.)

HOW TO MAKE THE VEGETABLE STONE,

OR QUINTESSENCE, FROM ALL GREEN HERBS, SEEDS, ROOTS, ETC., FROM WHICH THE WATER OF THE CLOUDS IS DISTILLED.

Now one will learn another method of distilling herbs, from which the water first comes over. Aside from that, one will understand all kids of herbs from which the water of the clouds is first distilled. (The right way of distillation is heating the herb solution and steam comes over and is cooled by cold running water — then the steam becomes (herb) brandy or alcohol which is inflammable).

For in the course of this operation and instruction relating to it, one will understand all that can be made of green herbs and roots. After that, the teaching will cover all dry species, gums, woods, and everything that is dry, each thing together with instructions relative to it.

CHAPTER I

First, my child, you must know that we intend to make the Vegetable Stone from green herbs, from which the cloud water is distilled. My child must therefore know above all at what time he is to gather and store the herbs, when they are strongest to make the stone thereof. Know then, my child, that the herbs have three periods: The first, when they are beginning to sprout. Then they are like a child when it first comes into the world, without strength or power, humid and watery. Likewise with herbs.

Their second period is like that of a 25 year old man. He is in his flowering until his 40th year. It is the same with herbs in their time of growth, till they begin to bloom and go into seed. Then they are in their flowering until the seed becomes ripe. The third period is like that of a 40 year old man till his 80th year, when all his forces begin to fall off. Likewise with herbs. When the seed is ripe, the herb begins gradually to pass away and wither, until it comes altogether to naught.

CHAPTER II

Therefore my child must pick the herbs when they are full-grown and the seed is beginning to come or to ripen. For all herbs go into seed and sometimes flower at the same time. That is why you must take the herbs that have mostly gone into seed, although a part of them are still flowering and have not yet gone to seed. Pick such together with leaves, flowers, roots, and seed, on a clear day, when the sun is shining strongest. Clean them without washing them or adding any moisture in your haste. Put them thus whole into a can, as thickly up to the brim as you can. Put the alembic gently on the can, and place the can into the Bath. Start distilling immediately, so that you do not lose the wild spirits that fly away invisibly. Of this I have taught in the Tractate of the Wine. The spirits are the greenness, the taste, and the smell, and their life. That is why the philosopher Dantin says: "Take care that you well preserve your greenness; otherwise you work in vain."

CHAPTER III

Until now, my child, I have taught at what time you must gather the herbs. Now we shall see from what herb we are going to make this Vegetable Stone. I do not find any ordinary herb less esteemed than *Chelidonia*[1]. I am telling you for sure, my child, that there are three herbs which have preference over all others. They are *Chelidonia, Solaria, and Lunaria*. All three serve the Art when they are prepared, and coagulate *Mercuris* into the true gold. About that you will be instructed in the Mineral

---

[1] greater celandine

Stone. I am telling you, my child, that the noblest of all three is *Chelidonia,* because the other two pass away in winter, while Chelidonia always remains in its greenness and flower. All other herbs of the world also wither and dry when it is very warm in the summer, but this Chelidonia always stays green. And even if it were lying under the snow throughout winter, it does not die. Therefore, it is not affected by heat or cold, dryness or wetness. It is the very best and strongest of the three.

GOD has infused such an influence into this herb that it cannot be sufficiently expressed by anyone. That is why, my child, we will draw the other and second Vegetable Stone from it, to cure all people of all diseases and let them spend their life in good health to the last hour, and at the same time coagulate *Mercurius* into fine gold.

We shall, therefore, pick this herb when it is in its first flowering. Clean it, and put as much as you are able to into three or four clean pots, without crushing it. Put the alembic on them immediately, and place them into the Bath in order to distill everything. Draw off all the water from it, till it is so dry that one could pulverize it. Then rub it on a stone with its water so that one could paint with it. Put it into a big stone pot. If you have filled four or five pots with *Chelidonia,* put everything together into a big stone pot. You must start with a great amount of herbs in order to obtain much *Materia* and much water. The uncrushed herbs take up much room.

CHAPTER IV

My child might ask, "Why do you not pound the herbs?" Know that if one were to pound the herbs, part of the three spirits would fly away, that is, the greenness, or color; a part of their delight, or taste; and a part of their natural warmth, since the three spirits are volatile that they cannot suffer any pounding or bruising. That is why you would lose the major part of them. And afterwards, your work would be Spoiled, because you would only operate on a dead body which would have been robbed of its soul and life, since the herb is *Mortified* by pounding it.

Try it: Pound a green herb very small in a mortar. It will quickly lose its green color and natural moisture, since the whole house is filled with the smell of the herb as it is being pounded. The smell, however, no longer grows when the nature of the herb has been broken. It is *mortified,* so that it is as it were estranged from Nature and the Influence of heaven which makes its fragrance grow. Nor do heaven or the stars, which give or throw their influence on it, give it any more help, because it has been broken off and therefore no longer gets help from any side. Therefore, the volatile spirits, which are its life, its soul, and its quintessence, part from it. Let the herbs, therefore, not suffer any pounding or crushing, as little as a man would like to be hacked into pieces; because the soul, which is his life, would immediately leave him. Consequently, my child, do not pound green herbs; but do as you have been instructed above so that you do not work on a dead body, as has been sufficiently proved above.

CHAPTER V

Now let us return to our work. You should rub everything that is left over a stone. Put it all together into a big pot, and the latter into a lukewarm Bath. Pour its own water upon it, and stir it well with a wooden spoon. Then put a small piece of cut glass on the mouth, and let it stand for two days and two nights. Stir it well every four or five hours, so that the water can well draw out the Elements.

At the end of the second day, take the pot out of the Bath and put it aside. Let it sink (settle) for three or four days. Then pour the liquid off the *fecibus* into a clean pot. Filter it, and pour the lic7uid into another pot. It is Golden Water. Cork it up and preserve it well.

Thereafter, pour some more water upon the *feces* and stir well. It would be good to dry the feces before pouring water upon them. Now put them back into the Bath for two days and two nights. Stir again, and cover. Then let everything get cold.

Do as before. Pour the water to the first one in the pot.

Take again of the water and pour it over the *feces*. Do this so often till the feces no longer color the water. Then you have the air and fire from the earth, and you have done enough watering. But should it happen that you have not got enough water from the herbs, you may take ordinary water from the Bath, distilled twice or three times, for all vegetable works, provided it is well distilled so that no *feces* remain. It is then just as good for adding to all green herbs.

Dry herbs, however, cannot be extracted or poured over with ordinary water but only with distilled *Aceto*. When, therefore, the water has thus been distilled from the herbs, take care of it, profit by it, and draw the elements out with ordinary water.

CHAPTER VI

And now, my child, we will go back to our work, to rectify our air and fire again together and to cleanse them from their *fecibus*. So, put all colored water into a clean dish. Now take the white of 40 or 50 eggs, beat it with a wooden spoon until it is thin like water. Pour it into the colored water, beat both together for a half hour so that they mix well. After this, put the kettle on the fire, let it become gradually hot and finally boil. But do not touch it at all. Now take it from the fire. Have at hand a big, white, woolen Ypocras-bag (sack). Pour all your water into it; let it trickle through into a glass pot. When it no longer drips, take distilled water and pour it into the sack upon the coagulated egg white. Let it sink through the *feces*, in order to draw the Elements from the fecibus. Do this as long as the feces give some color to the water. Then you have drawn all the Elements from them. Dry the feces in a pan and keep them. They must again be put into the Retort, to the earth, in order to draw from

them the combustible oil and the *Salarmeniac* for
there are many feces in the white of eggs.

Thereafter, take the liquid that has trickled
through the sack, put it into the Bath, distill it
so dry that it raises dust, and let it stand for 24
hours in the warm Bath. Stir it occasionally with a
wooden spoon, and again cover the mouth with a piece
of cut glass. Then take it out of the Bath and allow
it to settle down for three or four days. Now turn
it gently to one side and skim it carefully per
filtrum. Look if you can find a few feces at the
bottom. If not, it is sufficiently clear. If you do
find some feces, however, it is not clear and must
again be clarified as before.

CHAPTER VII

My child must know that all things in the world, if
one draws their water dry per distillatione so that
it raises dust, and one then pounds and boils it,
thereafter rubs it small on a stone, and again pours
the water on it which has been drawn from it, or
other ordinary distilled water, and one puts it into
the Bath, the water then draws to it all Elemental
water, air, and fire. It becomes red, and the
redness is contained in the innermost of the
greenness which the herb had. And as it coagulates

with the water and is dissolved, it leaves its feces each time. If this operation is often repeated, it will finally cleanse and purify itself until it leaves no more feces. Yet this is a long way. But with the water of eggs it is shorter.

But, as regards the green herbs which, after picking, are dried at the sun and pulverized, if you were to pour on them all the distilled water of the world, it would not extract or color anything. They must be drawn out with distilled vinegar. Neither will the vinegar turn red but a bad yellow, since their greenness is gone, which was their life, soul, and quintessence. The yellowness which it gets comes from the elements which are still in them, but the three spirits are mostly gone, and it is a dead corpse. Although it may still have something of the elements within itself, it is not worth while working on it. Therefore, mark well what I say.

CHAPTER VIII

Now we will take up our work again. When you do not find any more feces at the bottom of the pots, you must pour all the liquid together into a stone pot. Put it into the Bath and distill it down into water but not completely, so that you can pour the feces from the pot into a glass vessel; otherwise (if you were to distill ALL the water), you would have to break the pot. Thereafter, put the glass into a basin with sand above the basin filled with water; put therein the glass with the Materia. Let the water boil and the Material evaporate till it is dry. Then take the glass out, break it into pieces, and your Materia is clear, dry and red, and you have your Elemental water, fire and air. And you have your three spirits of which I have spoken be-for in a rectified and coagulated Massa - but not fixed. Put it in a dry room until we need it.

CHAPTER IX

In such a way, my child, the Elemental fire, the Elemental water, and the Elemental air, together with the three spirits, must all six be drawn into a mass from the earth, without distillation. They cannot be drawn out of the earth in any other way in the world, since the three previously mentioned spirits dwell in natural warmth and heat as taste, tincture, and smell.

These three cannot stand any heat coming from fire. If one would want to draw water, air and fire from the earth by distillation, it would have to be done through heat and dryness and through ashes or sand. The Elemental water, air and fire will not rise through the Bath, but solely the water of the clouds. If one would draw out the Elemental water, air and fire through distillation, it would have to be done without fire, since the three aforementioned spirits cannot stand the heat of the fire, which one applies with the fire. They would volatilize invisibly, and then you would be deprived of their soul, their life and their quintessence, and you would have a dead body. Nevertheless, you would have the four elements together, but they would be deprived of their soul and quintessence, which are keeping the four elements together and connect them. For when these three spirits are separated, the four elements cannot stay together but must part from each other. They begin to rot and die. Each element returns to its nature as air to air, fire to fire, water to water and earth to earth.

Take, for example, a man who has died and his natural warmth is gone. Very soon the color which dwelled in his blood, his natural smell and taste, all three are leaving the man; which three ARE the soul which ARE keeping the body together in one being. (Trans. Note: The version from Yale University Library has: "which three are keeping the soul together with the body in one being").

Understand well, my child, what it is when a child is conceived in the mother's womb by means of the natural help. Within 40 days a human being is thereby formed. All members are perfectly prepared by the warmth of nature which the mother has in her blood. For these three spirits are dwelling in the blood as in natural warmth, that is WARMTH (Trans. Note: Yale version has "color"), smell, and taste. From the blood of women the members of the child are formed with the help of nature, as it has pleased GOD. And thus these three spirits dwell in the blood of women, and the child's members are formed with the help of nature. Just so these three spirits are in all forms and members of the child, though very insignificantly; for the little members of a child, when its little parts are made within forty days, are at first so tender and small as if they were small wires, and it (the head) is like a small seed. Therefore, there cannot be much of those spirits in it.

As soon as the little members are formed in a minute way, God infuses the soul into them, which comes out of his supreme will, miraculously originating in it. We will not speak of that here, because it does not belong to this Materia. The soul has an eternal being, without the beginning, in GOD. That is why it comes out of GOD. And the little members have not been formed so minutely, the soul immediately comes into being and lives in the body. For if the soul did not immediately enter it, the three spirits would escape from it. That is why all three spirits must first be in the human being before the soul can dwell in it, and the three spirits are keeping the soul and the body together. As long as these three spirits are in the body, the soul also remains in it; and when the body becomes bigger, older and stronger, these three spirits also become gradually bigger, older and stronger. That is why they are called growing spirits. As soon as these three spirits leave the body, the soul must immediately follow and vacate it; for it has no spot or place where it can rest.

Try this in a man as soon as he is dead: Cut, do what you like, you will not find blood in it, neither heat nor warmth nor smell, but stench. Nevertheless, these four elements are in the body, e.g., the Elemental fire, air, water and earth, mixed with the stinking fecibus. But their quintessence is gone. That is, these three spirits, the natural warmth, the color and the air. GOD has adorned them - the elements - with these three and when these three spirits leave the four elements, they can no longer stay together but must separate, because they do not have the medium which keeps them all four together. Each goest to where it came from; nothing remains but stinking feces. And if one really knew this medium, all works would proceed more easily. But they do not notice that no spirit wants to stay together with the body without a medium which keeps spirit and body together. They do not know such a medium and do not know that the mediums must be spirits that are very volatile and lie in the depth of the Materia. It is an unknown spirit to the ignorant. More of it will be explained in the Mineral Stone.

Understand also, my child, what these three spirits are, since, if you do not know these three spirits and their nature, you will not make progress either in vegetable, animal or mineral matters, but will treat a dead body. That is why, for the reasons quoted, one can draw out these three spirits with fire and air, so that they coagulate together into a mass, but in no other way than the one we have taught. (The Yale text has: "...these three spirits cannot be drawn out before they stay together and coagulate into a mass"). Do not seek other means, or you will lose the three spirits invisibly and will then have a dead body. Understand my words thoroughly. They are open words and no parables, so that you should not be led into error.

CHAPTER X

Now we will return to our work. We take all feces
that have remained in the pot and the clarified
water of the eggs, where there are also some feces.
Put it into a big, earthenware Retort, well glazed
on the outside, as has been taught in the work of
the Wine. Put it into a furnace, in such a way that
the fire and the flame can get at it all around. Add
a large, stone pot, nearly full of distilled water,
to the neck of the Retort, lute it tightly. Give it
first a small fire, increased every three hours and
gradually stronger, for 24 hours, till the pot
(retort) heats through all around. Keep it standing
thus for six hours. Within that time the combustible
oil and the *Salarmeniac* will go over. Let it cool
down.

Now remove the pot (retort) and pour everything into
a large earthenware test (receptacle) that is well
glazed. Let it stand for three or four days. Now the
combustible oil will swim on the top. Remove it
carefully as neatly as possible. Then put the liquid
that is in the test in a large earthenware pot and
thus keep it until such time as you must rectify it
by coagulating and dissolving it.

Take the combustible oil and put it into the little vessel (barrel) about which I have taught in the work of the Wine. Pour distilled water, boiling, on it and start churning as if you were to make butter. This is as has been taught in the work of the Wine, where the combustible oil is cleansed of the Salarmeniac. It is all one operation.

When the oil is cleansed, put it into a clean glass. Use it. It serves for all sufferings that come from cold and humid diseases, to anoint all lame members, and in the paralysis. After that, take the water in which the combustible oil has been purified, and the water from which the combustible oil has been skimmed off. Put everything together in a Bath so as to coagulate and to let the feces settle down and be drawn off by filtration, as has been taught in the work of the Wine, to rectify of the Salarmeniac. When your Salarmeniac is well rectified and also dry, as white as snow, keep it in a very dry place.

CHAPTER XI

After this, take all the feces that have remained in the Retort, also those left during the rectification. Put them all together to reverberate, as has been taught in the work of the Wine, until they become snow white. Then rectify them again with distilled water by pouring it over them and letting it stand over them. Afterwards, let the feces sink down, and then draw them off per filtrum, and again coagulate them. Do this as has been taught in the work of the Wine, until your earth is white as snow.

Now take the white earth, dissolve it in your rectified water. Put your Salmiac into the same water; draw the water off until it is as dry that it draws dust. Then put it into the egg, to calcinate in the secret furnace, and do as above. When all has been calcinated, dissolve it in your Aqua rectificata. Let the feces sink, draw them off by filtration; coagulate, and do as before till no more feces remain. After that, coagulate again your Elemental water, fire, air and earth. Then you have rectified your mass of the outer and inner fecibus, and also your Salmiac. They are now prepared to make the Vegetable Stone from them.

CHAPTER XII

Following that, take a big receptacle, as has been taught in the work of the Wine. Put therein your Salarmeniac, Elemental water, air, fire and earth, together with their three spirits. Pour upon them some of your rectified water, which has been drawn from them, so that it may dissolve correctly, and no more. Now put it into a crucible with strained ashes. Cover the glass with a cut little glass, unglazed, and a weight on top of it. Give it heat like the Sun in the middle of the Summer for twenty-four days.

Then let it cool down, pour it into the egg, and put it into a crucible with sifted ashes. Let it evaporate in a gentle heat till everything is quite dry, which you should test with a sharp knife put on the mouth of the eye-neck. Look if there is steam forming on it. If no moisture forms on it, it is dry. But, in order to be more certain, let it stand for three or four days in the warmth. Following that, fixate it with the Lute of Hermes, and hang it into the secret furnace for 40 days. Heat it like the sun shines in the summer, or somewhat hotter. After the 40th day, let it cool down.

Break the glass, take the powder out, put it into a glass crucible of Venetian glass. Place it on hot coals. The powder will melt like wax. Pour it into a small glass form, previously greased with oil. When it is cold, it becomes hard like a stone, clear like a crystal, red like a ruby, transparent. This is the second vegetable stone, which cures all diseases and infirmities of the world. If one takes every day in

wine as much as a grain of wheat is heavy, you will
see wonder upon wonder in a few days.

CHAPTER XIII

Furthermore, if you wish to achieve that it (the
stone) coagulate Mercurius into the true gold, pound
your stone into powder, and put it into a very thick
glass. Then take fine gold, which has been cemented
and dissolved in Aquafort, which must be made of
equal parts of Salarmeniac and saltpeter. Dissolve
as much gold in it as your stone weighs. When it is
dissolved, distill the Aquafort dry, and prepare the
gold lime, so that it may dissolve in good Aguavit.
Then it will tincture beautifully yellow.

Pour off what is clear, and pour again more Aquavit
upon it till there is no more tincture. Put it away.
What stays at the bottom is salt from the strong
water which is not dissolved in rectified Aquavitae.
Then evaporate your Aquavitae from the gold;
dissolve it again with fresh Aquavitae. Pour it off
from the fecibus, and continue in this way till no
more feces remain at the bottom. Keep the feces
somewhere.

Dissolve and coagulate the gold till it turns into
an oil which will no longer coagulate. Then it is

prepared. Or when it is first dissolved in the Aquafort, as said before, pour upon it a large amount of fresh, ordinary water. Put your glass over the fire, let it boil for one hour, then put it to one side, and let it settle into a powder for three or four days.

Draw liquid off by inclining glass to one side, or by filtration. After that, pour it off and dry your powder in a glass dish, on warm ashes. When it is dry, put it into a glass such as you see here. Pour well rectified Aquavitae upon it, place it in a crucible with sifted ashes, close the mouth of the glass with a cork or put a small head upon it. Put a receptacle in the spout, glaze, and heat it like the sun shines in the summer. Then your gold will nicely dissolve. The amount of the Aquavita which rises above it, pour again upon the gold through the button (knob) of the head in which there should be a hole and a little glass funnel. Let it stand for 8 or 10 days, and your Aquavita will become nicely golden-yellow.

But if everything has not yet been dissolved, so that there stays powder at the bottom, continue pouring off from above and fresh Aquavitae upon the powder. Put the head on again and do as before till all your gold is dissolved in the Aquavitae. Then take your dissolved gold together with the Aquavitae and pour it upon the vegetable stone which you have pulverized. Put it in a glass pot; put the heat on the pot; put the pot into a crucible with strained ashes and give it a gentle heat, like that of the sun shining in the summer. Then your powder of the stone will be dissolved in Aquavitae with your dissolved gold. When you see that everything is dissolved, give it a bit stronger fire, so that the Aquavitae is distilled off. That will be a slow process, because the stone and the gold coagulate the Aquavitae in themselves and keep it.

When you have drawn some off, pour it back on again through the heat, with a glass funnel; fix the receptacle back on it, glaze and distill again. Each time you take it off, you must increase the fire; for the more you pour on, the more stays with the stone, so that it will no longer come over. Then the Stone is fixed, sweet, and delicious. It has converted the Aquavitae into its nature with the gold. As long as it is warm, pour the stone into a small form. It will immediately coagulate.

After that, you must again pulverize it and put it into the egg, sealed. Hang it into the secret furnace for 21 or 31 days. Put fire underneath it like the sun shining in March. You must not give it more heat, because it would melt in the fire, since it is fusible in a little bit of fire. If it were to melt in the egg, the gold in the Aquavitae would be calcinated or distilled together with it (The Yale text says: "...the gold and Aqua vitae would not be calcinated and distilled with it (the stone) for there must be nothing in the stone that is not calcinated, and yet the gold and the Aqua vitae would not be calcinated.")

Then (at the end of the 31 days), take the egg out and break it. Take a big glass, put the powder of your stone inside; pour a large amount of your rectified ordinary water upon it; put it into the ashes or the Bath, it does not matter which. As soon as it is warm, it will dissolve. When it is dissolved, immediately put the fire in the furnace out. Let the glass stand in a crucible, and the feces from both the Aquavitae and the gold will sink down. For there is nothing in the world so pure that it would not have a combustible oil and impure feces in its innermost nature. And that cannot be purged out of it, unless its body be first mortified and die, meaning, that it cannot again become a body. After that it must be calcinated in a secret furnace, each according to its nature; one with more the other with less heat; the third with a gentle, the fourth with a lukewarm fire, before its innermost feces can be drawn out. And in the stone there must not be any feces at all, either inside or outside. That is why the gold must be put into the secret furnace together with the Aquavitae in order to be calcinated, if one is to bring out their innermost feces.

In this way the feces are drawn from the gold and separated from its body and have sunk to the bottom of the glass. When it has thus stood for four or five days in order to sink, drain it carefully through filtration into another glass. Again, pour more rectified Aquavitae upon the feces that remain in the glass, stir well, let them sink, and again drain them as before. Continue doing this till the feces no longer color the water. Then you have all the power of the stone out of the fecibus.

Coagulate your stone upon warm ashes in an open vessel. Or if you wish to keep the water, drain it with a head. When it (the stone) is dry, melt it and pour it into a small glass form. Thus your stone is ready to coagulate Mercurius into fine gold.

My child must know that CHELIDONIA thus prepared will coagulate and fix Mercurius into real gold. The same for SOLARIA and LUNARIA if they are prepared in this and no other way.

Do you believe that the art lies in herbs or other things (except in gold and silver?) Don't you let such thoughts arise in you, or you, together with many fools, will be greatly mistaken. Do not seek in a thing what is not in it, of which I will teach more in the Mineral stone.

My child must know that in this way one can make a vegetable stone from all green herbs. In addition, there will be instructions concerning the making of another Vegetable Stone, to heal all diseases of the world. And the third manner of operation proceeds from sugar, because out of that a noble Vegetable Stone arises.

OPERA VEGETABILIA

by
Johannes Isaacus Hollandus

PREFACE OF HOLLANDUS

My child should know in the course of time that God
the Almighty Lord has created heaven and everything
in it, and the world and everything in it, as is
written in the Book of Genesis, and that the first
Materia was water, upon which the Spirit of the Lord
rested. Therefore I say, my child, *principaliter,*
that nothing in the world is naturally composed by
God the Lord *substantially* out of the four elements.
In his first point of creation, it is generated
substantialiter from *Sulphur* and *Mercurius,* pure and
clean and incombustible. If then, my child, all
things of this world have their substantial, special
and accidental form out of the first Materia, it
follows clearly that there is nothing in the world,
no matter which, that is not originally and
principaliter composed of the said *Materiae* Sulphur
and Mercurius. And when nature wants to give birth
to anything into a *substantial* form, it takes the
first *Materia,* which is then still a simple and
imperfect form, and begins to fix in it the four
elements, which are of varied natures. According to
the difference in the mixture of those elements, and

their purity or impurity, different *complexions* arise and different figures; also different smells, colors and tastes in the *mineral* as well as the *vegetable and animal* (things). Since the first—mentioned Materia is simple and *uncomplexioned* (not put together into one complex matter), it may assume an infinite number of various forms. After the four elements become *complexioned,* this way or that, it also has the power to move from one form into another. If then they get good nourishment, pure and clean, they will assume a noble form and a delightly quality, which will be noble and good.

Further, when God the Lord had created all things in that way, he infused into them five common natures, so that all human beings, cattle, fish and all other animals, yes, trees, herbs, plants, and whatever God has created in the world, all have five natures implanted into them. One is the nature of *generation,* i.e., that each thing should generate its like and not otherwise. Man shall generate a man, and not otherwise. For God has not given to man to beget anything but a man, since he cannot give what he has not. What is not in a thing, you cannot take out of it. The same applies to all other animals. A horse begets a horse; a fox, a fox; a fish, a fish, etc. The same may be noticed in all herbs, and in everything God has created. No apple -

or pear tree can grow from a turnip seed. Each begets its like, as said, for what is not in a thing cannot be brought out of it. All things, therefore, have a common nature of generation, each in itself and its like.

Aside from this, God has given two further natures to what he has created and made. One is an active nature, the other a growing nature. These two natures cannot be one without the other, because the two natures must help each other, as will be clearly proven and taught at the end of the book. When the active nature begins to act upon a Materia, or just created things, the growing nature must also be present. If it were not immediately ready with its nourishment, the active nature could not operate long on a thing. It would immediately spoil and come to naught. For example, when the male semen reaches the *matrix* of the woman and nature would soon begin to work - which she does, provided she is not hindered by other things — and when then nature begins to work, the growing nature must immediately be ready with its nourishment. These two natures do not stop but are both at work without ceasing until they come to the time when they reach the end to which GOD has ordained them by giving them a certain measure above which they cannot go. When those two natures have reached the same appointed time and the specific measure given and set to them by GOD the LORD, they stop their working and growing as long as they are not moved further to work more.

"When does it happen, however, that these two natures omit their work in a created thing?" my child would like to ask. My child should know that GOD the LORD has given a certain size to every man, animal, bird, fish, tree, herb, plant, and everything created by God, provided it is not hindered by haphazard accidents or by plagues of God, and a certain fixed number of years beyond which it cannot go. As we may see, one becomes big, the other, small; one large, the other narrow; one pretty, the other ugly. Such is caused by way of the elements and the two-fold influence of the planets and the fixed stars, all of which are active in them according to the mingling of the elements and the said influence of heaven. After that, then, a thing is strong or weak, beautiful or loathsome, small or big, or it lives long or a short time. This will later on be better explained when the mixing of the elements is taught and the elemental nature infused into them by God will be clearly proven.

Thus, my child, when these two natures have brought a thing to its perfect size and its perfect power, if it is not hindered by something else, these two natures separate again from it and are no longer active in it because they are no longer moved (impelled). God the Lord, however, has not made anything here on earth or up above in the sky that can stand still. It must be moved either for good or for evil, for rising or for declining. Therefore, when the first two natures have done the work for which they were moved, when they have brought the created thing to its whole strength and perfect size and thickness, that is, to its highest power to which God had ordained it, the two natures cannot act further in it, because they are not moved further to (produce) more work and nourishment.

From then on, they separate from the created thing. Yet no created thing can stand still but must necessarily be moved, as will be taught later. Thereupon, immediately, at the same moment, yes a thousand times faster than the first two natures which have brought the created thing to its highest perfection, leave it, so that here on earth it can never again attain more strength and a greater size through the action of nature. Understand me correctly, my child! I except the Art, because a thing may well attain more power through the Art, but here I simply speak of the action and nourishment of nature. These two cannot bring the created thing to greater size and strength, or to more loftiness, more nobility, because they are not stirred further to great power. Thus the two part from it, and immediately the last two natures arrive as the sick-making, or declining, or decreasing, or going—backward nature, and the suffering nature.

Those are two natures which remain with the created thing till it again attains its first Materia of which it was composed and created by God in the beginning; but not to the same degree, because a human being does not again turn into a human sperm; neither does a herb to a small seed from which the herb grew up in the first place. A big apple or pear tree will not again become a pip out of which it had sprouted in the beginning, but it will look as though it were again coming to naught. That, however, will occur to a different degree, as will be clearly proven later, when we shall write of the glorification which God will accomplish in all created things at the Last Judgment in this very place. Likewise, when we shall treat of the perfection of the stone, as much in the vegetable as in the animal or mineral, where it will be clearly taught. For everything said in this preface of the first Materia and of the different natures which God has created and incorporated in these lower things, and other *rationes,* will later be gathered in order to attain to the *proposito,* understanding and

complete comprehension of our Art, as also to the
perfection of the three stones. For I am telling
you, my child, if we do not know the first Materia
of a created thing and its nature, beginning, middle
and end, inside and outside, all its infirmities,
all its circumstances, and all its powers, as also
everything that may get or fall into it, we shall
work as a blind man shooting at a little bird. That
is why my child must get to know the first Materia
of all created things and their nature, as also all
their powers and sicknesses.

But we will now return to our two natures, being the
sick—making and suffering natures. These two work in
a created thing to the contrary of the first two
natures, of which I first reported. For the first
two, being acting and nourishing or growing, were
working and nourishing gradually by degrees, for a
long time, until the created thing had reached its
highest perfection, so that the said two natures
could no longer work in it, since they were not
called to accomplish any further perfection. Then
those two had necessarily to retreat from the
created thing. Now the created thing could not stand
still, as I informed you, but had to be moved by one
or the other nature. Therefore, there must (needs)
be two other natures in the created thing, since the
sick-making or decreasing nature, and the decreasing

or sick-making nature could not be alone in the created thing but must have a companion with it, being the suffering nature. For the one, being the sick-making nature in the created thing consists in working; and if it had not also found a suffering nature, it could in no way destroy the thing with its sickness. For if there had not been a suffering nature in it, what would the sickness have acted upon? Likewise, as the acting nature consisted in the beginning in working, if it had not had a companion in the leading nature, upon what would the acting nature have worked?

My child, understand thoroughly the words I am teaching you here, for it is the root from which all natural arts arise, as you will probably later understand better when it will be clearly proven. Thus, these effects must take place equally in the three stones, in the same way as in our works. If they are to reach their highest potency, these first five natures must first and above all happen (ereignen) or unite. When that has been done, our stone will attain such power and strength and such a perfection that no nature will have the power to act in it, but it (our stone) will have the power to act in all other natures, to push out all foreign natures from a created thing, and to drive them away, and to bring the created thing to its nature.

Understand well, therefore, what I write about those natures, because that must be well absorbed in the first place, as one would get a wrong basic understanding when all previously told things will later be dealt with in detail and clearly. Learn, therefore, my child, to understand thoroughly the matters now discussed, then you will afterwards understand them all the better.

Therefore, just as the two aforementioned natures have brought the created thing to completion *gradatim,* over a long time, just so the last two natures have also worked in the created thing, by degrees, gradually, over a long time, until they have brought the created thing to the same *termin* (end or state) in which it was when it was first created in its creation, but not to the same degree at which the first two began to work.

My child, do well understand the meaning (of my words), because this discourse is somewhat difficult to understand. If I were personally present with you, I would explain it more clearly and understandably. These last two natures of which we have just spoken, are to be used from the end to the first work of all stones, in order to operate back (to undo) that which nature has worked of itself within them, until we have made it spiritual again;

so that the Art must again undo what nature has done in them. Although the Art cannot work in the created thing to again undo, as I have taught, that which the last two natures have broken down after the first two natures had built it up. For the last two natures have gradually, by degrees and over a long time, chased away and driven out the three spirits, or the quintessence, from the created thing, so that they must finally leave the created thing totally. When the three spirits are out of it, those two, the sick-making and the suffering natures, get two companions to help them, that is, the fat salts or the combustible oil, and ✳ . And immediately these four separate and destroy the whole mixture, and it will remain separated until Judgment Day, when God will *repair* them again to their perfection.

True, the Art must work in a created thing in order to undo what nature has done; but the Art must work in the created thing *in contrarium* to what the last two natures have done in it. For the last two natures have robbed the whole mixture of its spirits or its quintessence. They have caused the whole *massa* of the mixture to separate from one another, and have destroyed it. Against that the Art must to the *contrarium,* and again undo in the created thing what the first two natures did. But it must preserve the three spirits and the four elements with their *salts* in such a way that they are not diminished in the least. It must make the created thing spiritual, as it was before, but not to the same degree. Thus, my child, we must follow nature in some works; in some other works, however, against nature, we must operate quite the contrary, as is proven here. Even if not everything is pertinent, I am writing this for my child, so that my child may better understand my views in this preface which lain here writing about the first *Materia* and the natures, to enable you by greatest diligence to comprehend my words completely.

Thus I have now indicated to my child the first *materia* out of which such *prima materia* is composed, and what it is, including what parts it has. Aside from this, my child should also understand that, although the first materia is (composed) of many parts, it is nevertheless only one part; for one part cannot be without the other, and (the whole) is therefore no more than one part. Just as there are three Persons, as Father, Son and Holy Spirit, it is yet but one God, for one cannot be without the other. Thus and not otherwise is it also with the first *materia,* as will be clearly proven hereafter.

Above all you must know that, if you wish to try your hands at the practice, you must recognize four things, my child. And unless my child knows these four things, you will not accomplish anything. First, my child must know a dead *corpus* from a living body. In addition, you must know a complete *corpus,* meaning: A body which is in its full power, on which the aforementioned two natures have accomplished all their work, so that it is in full power and its highest strength and force. The third knowledge is that my child must know of what natures the corpus is on which you wish to operate, whether it contains that which you wish to extract; for one cannot draw out of anything that which is not in it. For if you wished to draw sweet wine from a cask of vinegar, it would be impossible, because nothing can give what it has not itself. The fourth knowledge is that my child must know a *simple corpus* from a *composito,* so that my child does not mistake a *corpus compositum* for a simple corpus. For if my child were to take a composent (composed) corpus for a simple corpus, my child would make a mistake.

Here ends the *Prologue,* in which *Theoretica* has been lightly touched upon, aside from many references that are important for the understanding of the philosophical Art, which is secret. And therefore I advise you who will read this theory, that you do not reveal it to anyone, unless they are true lovers of the Art. But if you do otherwise, you act like a simpleton and fool, for it would sound strange in the ears of the ignorant.

THE FIRST PART OF THE VEGETABLE STONE OF THE WINE

BY ISAAC HOLLANDUS

CHAPTER I

In order to obtain such a medicine which cures and removes all sicknesses and keeps the healthy, healthy and drives away old age, while keeping it in good condition to the last *termin* of life, as has been set by God the Almighty, it is first and above everything else necessary to know the four elements and their nature, inside and outside, their power and their *feces,* and what is contained in them. For in them there are two natures, one that is perishable and eternal. Therefore it is necessary that we should first have a knowledge of the elements, since everything in the world has originated in and is composed of the four elements. Accordingly, my child should know that I wish to teach him, in undisguised words and without the interference of foreign termini, the right truth to reach this Art and wondrous medicine.

CHAPTER II

My child should know that the divine medicine consists in three types of knowledge and three types of work by the hands. They have three special names, although they are nevertheless all one, just as the Holy Trinity consists of three separate Persons with three special names, such as, Father, Son and Holy Ghost; (but) the Father is not the Son, and the Son is not the Father or the Holy Ghost; and the Holy Ghost is neither the Father nor the Son. Yet each is a separate Person, and nevertheless these three separate Persons are one being and have one might and one power. Thus it is also with these three different medicines; one is not the other, but although they are of three different kinds, they are nevertheless one in essence and of the same nature and power, but with three different names: Vegetable, Animal, and Mineral.

By "Vegetable" you must understand everything growing out of the soil, such as herbs, trees, spices, fruits; and everything sprouting out of the soil, such as roots, flowers, etc. The second is called "Animal", which refers to everything that has life and feeling in it, such as human beings, animals, cattle, birds, fish, worms, and everything that has received life. The name of the third is "Mineral", namely everything that grows in the earth, such as gold, silver, and all metals, minerals, marcasites, rocks, and everything that comes out of mines. These are three separate names and three separate natures and beings, and differentiated in three separate substances. Yet when they are brought to their highest potency to which God has created them, they are one in nature and retain the same equal power and being in all eternity, as will be taught later (God willing!).

CHAPTER III

Above all, my child should know that water was created first, and to that water God incorporated his earth. Out of the earth all things have sprouted, and out of it everything has grown that has received its being and life by the will of God, with the help of the upper choirs, such as the sun, moon, planets and stars, which together pour their influence and power over them, in the way God has provided in his divine order, as will be discussed subsequently.

CHAPTER IV

Further, there are two manifest elements, such as water and earth, in which two others are concealed, namely, air and fire, which are *influencing* elements. Air is contained in water, and fire in earth; and they are so knit together that they can never be rightly separated. In addition, earth and fire are fixed, whereas the other two, water and air, are volatile. That is why water rises together with air, and earth and fire stay together at the bottom. Among these four elements, fire and water are opposed to each other, as are earth and air. But air *symphonizes,* or equalizes, with fire in warmth, and with water in humidity. Likewise water symphonizes with earth in coldness. Earth has equality with fire in dryness. Which explains clearly that each element can be made *concordant* with two other elements, and that in property one is contrary or opposite the other.

CHAPTER V

Further, I admonish my child to know that there are two kinds of beings in every element: One is perishable, subject to decay, corruptible and combustible; the other, however, is eternal and imperishable like the indestructible heaven, also of a heavenly nature, so that it can neither rot nor be burnt by fire.

In addition, in these two natures there is still another one concealed and mixed with them. It is called: Rotten, stinking *feces.* It is so much united and mixed with them that it robs them of all their power, so that they have little or no power; and it makes the elements stinking and *putrefactable,* of which we are now going to speak.

CHAPTER VI

My child should know that there are two kinds of
water. One is the water of the clouds, a lake or a
creek; the other is the *Elemental* water, and this
latter counts as the element water. It is the water
of the philosophers, which the ignorant do not know.
Likewise, there is an earth which is white, pure,
shining and eternal. It is the earth of the
philosophers. Aside from this, there is another,
black, stinking and combustible earth. In the same
way, there is an *Elemental* fire, which is eternal
and is the fire of the philosophers; against that,
there is another fire which is stinking and
combustible. Likewise there is an air which is
elemental and is the air of the philosophers. In
addition, there is also a stinking and combustible
air. These base things are mixed with the rest and
are the reason why all things in the world are so
easily destroyed, so that nothing can last long, but
they bring them death and corrupt all natures, by
and by, no matter how noble they may be. This is
true for all things, vegetable, mineral and animal.

CHAPTER VII

Now I will teach my child how he is to separate, by
Art, the eternal, elemental nature from this

corruptible nature. I will also inform him of what
harm it can bring to human beings, cattle and
animals. However, to obtain such a thing that is
harmless to nature, necessity demands to learn above
everything the manner in which we must separate the
eternal from the perishable, stinking and
combustible. After this, it serves my child to know
that many mistakes are made in the separation of the
elements, inasmuch as some ignorant people are found
who put herbs and *species* (spices) to putrefy,
pretending that they intend to separate the elements
from them in the following way:

When they take them out of putrefaction, they draw
from them the water *per balneum* or by fire. Then
they remove the *materia* from which the water has
been distilled and rub it with its own water. Now,
they distill it again by fire. Then a yellow water
goes over which they separate *per balneum.* The water
becomes clear and the yellow matter stays *in fundo*
and will not rise in the *balneum.* This they say to
be the element air; but they do not know what they
are saying, since it is some impurity of the fire
and the air which had risen with the water during
distillation because of the stronger burning of the
fire used for the distillation.

Thereafter they take what has remained in the pot and rub it with its own water. Then they put it into putrefaction for 6 or 7 days, and afterwards distill it again on fire. Then a red, thick and fat *materia* goes over, which they call the element fire. They remove it from the *balneum* (or: they separate it in the balneum), and the red, fat *materia* stays at the bottom and will not get out of the *balneum.* They are taking it out and then say that they have the element fire. Now they calcinate what remained at the bottom of the pot and draw the salt out of the *fecibus.* After that, they rectify each element by itself and are heard to say that they have separated the four elements, although they do not know what they are saying or doing; neither do they understand themselves nor the work they have done; nor do they have a knowledge of the elements.

The proof is as follows: First, they set to putrefy and rot a thing, when the external heat in which it stands drives the inner, natural warmth out of the thing they putrefy. For the natural warmth a thing has in it is a spirit with which three kinds of spirits are mingled. One of them is the *color* of the thing, be it an herb, a flower, or a spice. That is its green, red, brown, yellow or other color which an herb, foliage or flower has. The second spirit is the *taste,* and the third spirit is the *smell* or air which everything gives off, and this (smell) is the subtlest of all three spirits. Therefore, these three are subtle spirits which escape so adroitly and unnoticeably that the philosophers have therefore called them wild spirits, which cannot be fixed, although the Artist can nevertheless fix them with skill in the work. These three spirits take their beginning in everything, be it a vegetable — or animal *corpus*. As soon as that thing has received its power or form, it is ready to receive these three spirits, by the will and decree of God, which he has infused into nature, each thing according to its kind.

These three spirits grow up gradually with the body, and the bigger, larger and stronger the body becomes the bigger, larger and stronger these three spirits become, each in its own powers, namely: In taste, smell and color until the thing reaches its highest power, that is, when it is fully grown. Then, my child, the herbs should be picked, when they are fully grown and not when they are half—grown; nor when they are withered, since they are then in (the state of) decreasing in strength.

CHAPTER VIII

After a thing has reached its greatest growth, it is in its highest potency. Subsequently, it begins to

110

go down and to decrease in taste, smell and color, until it has come to naught. Finally, these three spirits escape completely from it. As soon as they are out of it, the thing (be it vegetable or animal) is dead, becomes evil—smelling and decays. The same occurs during putrefaction. What they set into it may well be alive and good, but when they take it out again, they are rid of these three spirits and the matter is dead, stinking, and rots.

CHAPTER IX

Yet none of these three spirits is of the *Elemental* elements, but God has adorned and clothed the elements with these three spirits, and they are their life and soul. Of them, the Fire is animal, the Water, Aire elementall and no man but God can separate them but the water of the clouds may be separated from them. Also, all the feces may be separated from them, which are mingled with them; which are the stinking and corruptible elements, and the four elemental—elements may be brought to a chrystalline shining. But these three elements, Fire, Air and Earth are unseparable.

Then they distill this in the *Balneum* and withdraw its own water, and rub that which stays at the bottom with its own water. Then they set it again to putrefy in manure or in the *Balneum,* for 7 or 8 days, maybe also 10 or more; so that, if anything good of these three spirits should have remained, they drive it away completely and are thus spoiling it in one go. To this they may now object: We lute the vessels quite firmly before we put them into putrefaction. Know then, my child, that if a glass were a foot thick and were ever so strongly hermetically luted, it would nevertheless break into a hundred pieces if the putrefaction were to get heated and the spirits were to rise. Try it, you will find that it is so. Well then, they lute with some materials of *lutaments,* upon which they put their trust. But, although they were to lute a foot thick with the strongest lutaments one could find, the three spirits are yet so subtle that they penetrate invisibly. Try it, put fragrant herbs for 40 days into putrefaction, and when you take them out, all their natural fragrance is gone; they smell sour and stink, no matter how strongly you may have luted.

CHAPTER X

The other reason is: When then they take it (the matter) out of putrefaction, they put it to distill in a vessel set in sand or ashes, so that a yellow water goes over. They call it the element air. It rises over together with the water. However, they do not know what they are saying, inasmuch as they are not aware that the element air cannot be separated from the element water, although one can well separate the water from the earth. But the *Elemental*-water, the *Elemental*-air, and the *Elemental*-fire, these three cannot be separated by anyone in the world but by God alone, who has the power, and no one else; while they are united and married that they cannot be separated either now or in all eternity.

True, one may well separate from them the water of the clouds, which is moist and running water and, in addition, all corruption and *feces* mixed with it, so that one can bring the elements to crystal clearness. But the three elements, air, water and fire cannot be separated. The ignorant cannot understand that there can be no fire without air; for if the air has been drawn from the fire, the heat of the fire would have to choke, die, and come to naught. And if the fire were drawn away from the air, it would be mere water; because the air is warm and humid and *participates* with water and fire. If then the fire had been drawn from it, it would be all water. And if water were drawn from the air, it would be all fire. Thus you can easily understand that the said three elements are inseparable.

One may, of course, separate those three elements from the earth, but not completely. Some earth must remain with them; otherwise one could not make a *corpus* with them which would be tangible *in specie*. That is why the elements cannot be separated, because the three elements fire, air and water carry with them, from the gross part of the earth, a subtle *terrestrial portion*. They incorporate themselves with it, so that they become dry and tangible but not fixed. For if one wishes to fix them, they must possess something of the gross part of the earth. Yet the Master must render the gross parts of the earth subtle before putting them together. How this can be done will be taught later.

The reason why the water goes over yellow, is that it is distilled on fire, and that the burning of the fire drives up part of the air and the fire together with the water, which causes the yellow coloring. Then they put it into the *Balneum* and *abstract* the water, while the air and the fire remain *in fundo* of the vessel, together with many *feces* which they preserve. Further, they pound that which first stayed back at the bottom of the vessel and rub or *imbibe* it with its water. They then put it again into putrefaction, for 8 or 10 days, according to their foolish whim. Afterwards, they put it on the furnace, and distill by fire, gently to begin with and subsequently by a stronger fire. In so doing, they drive out everything they can, so that the vessel with the *materia* starts to glow. Then they say that the fire has gone over together with the water. But they do not know what they are saying; nor do they know that which goes over. These are the reasons why: Because all the other three elements went over with the water, that is, the air, the water, the fire, and a part of the earth which went partly over with the other elements on account of the strong fire. What also wandered across was the combustible oil and the ✳, which two are likewise concealed in the elements and intermixed with them.

CHAPTER XI

For just as the three noble spirits are hidden in the elements, these three, that is, the combustible oil, the **?** (symbol missing) and the impure *feces* are likewise hidden in the elements. The first three, however, are so volatile that they go over first or separate first from the elements; and the last three separate last from the elements.

CHAPTER XII

That is why my child should be aware that the first *materia* of everything in the world was *Mercurius;*

since water was before time was, and the Spirit of the Lord rested on the water. But what kind of water was it? Was it water of the Clouds? Or a moisture that could be poured out? No, but it was a dry water unto which God hath joined His earth which was his *Sulphur,* so that the earth coagulated with the water. And out of it came the four elements which were ordained in these two by the command of God and His supreme will. *Mercury* dissolves the *Sulphur,* and *Sulphur* coagulates *Mercury.*

And these two cannot be one with the other, for *Mercury* is never without the *Sulphur,* whereas it is being transformed into it. For the nature proper of *Mercury* is that it dissolves its *Sulphur* and whitens it; and the nature of the dry *Sulphur* purges and coagulates its *Mercury.* And as these two cannot be one without the other, they cannot be without *Salt,* which is the principal means whereby nature accomplishes all her *generation* in all things, in vegetable as well as mineral and animal (works). May you well understand my words!

For if nature did not have *Mercury* in her generation, straight at the beginning of the original composition of every created thing, it could not keep together in natural humidity, which is one of the most necessary items for keeping a thing in its essence. And if she did not have *Sulphur,* the humid parts could not be coagulated. In the same way, if she did not have *Salt* (a means which connects both and causes one to enter the other), it would not mix or unite with anything in the world; because there would be no sharpness to penetrate, or it could not mix with anything. Therefore these three, *Mercury, Sulphur* and *Salt* do not exist one without the other. Where you find one of them, you find all three; and there is no created thing in the world wherein you do not find them. And from these three, everything in the world has sprung. They are also in the four elements, mingled in such a way that they are one in one body. *Salt,* however, hides in the very deepest of the elements, which it must keep them together with its sharpness and dryness. Nevertheless, it is a spirit and volatile.

However, because it is contained in the deepest of the mixture and is kept under by the fat combustible oil to which it clings — for the salt lies in the combustible oil like the yolk in the egg, and the combustible oil lies in the deepest of the elements whence, together with the *Salt* and the *fecibus,* it separates last from the earth, and the salt lies buried at the bottom of the feces of the earth and the combustible oil - it cannot flee from the earth except by the power of the fire.

These three Spirits, whereof we have spoken must first be separated from the mixture of the elements, which is the soul of all things, or their *Quintessence*. This is what binds together the whole mixture of the elements. For when the spirits are drawn out, then the mixture will dissolve or separate of its own and is divided. Neither is there need of fire to expel the salt from the earth; when the elements are separated from the feces, then is the salt also separated with it. This salt is unknown to the ignorant, because it is contained in the deepest of the elements. Those, therefore, who do not know this salt, must remain in error. Salt therefore, is the means between the gross, earthly parts, and the three volatile spirits resting in the natural heat. That is, the taste, moist-smell and color. These three are the life, soul and quintessence of everything nor can one of these three be one without the others.

CHAPTER XIII

Previously, I told you that these three spirits which escape invisibly during putrefaction are so subtle, that they cannot be fixed. Whereupon I immediately said: They can nevertheless be fixed together with the coarse part of the earth when the same has been rendered subtle by a Master, Who is skillful in the Work. And with this Salt, which is a medium between these coarse, fixed parts of the earth and the three volatile spirits, these Spirits are like unto the Father, Son and Holy Ghost, being one, yet three persons, and one not being without the others. Which is why they are the life and soul of all created things. The *Quintessence.*

But the ignorant understand it not and make fun thereof. For these three spirits, being tied to the gross and earthly fixed parts, if they are subtilized and if their Salt is joined and mingled with them, the one penetrates the other and fixes them into a crystalline body which is Diaphanous, red and transparent like a ruby, whereof we will instruct you later. But those who do not know this Salt, they will never achieve anything in the Art. Philosophers have called this Salt a dry water and a lively salt. But the ignorant thought they meant Mercury thereby. They also named it an ensouled Salt and concealed its name. They also called the three spirits Mercury and gave Earth the name of Sulphur which the ignorant also did not understand. Now one cannot be without the other and there is no created thing in the world wherein these are not all together, yet so intertwined with the four elements that they form one *massa* (mixture) or *corpus* (body).

CHAPTER XIV

So, I am saying once again that those who distill in that way and drive one thing over with the other in the fire, they do not themselves know what they are doing. Neither have they noticed that all their work and distillations are stupid business; because they know not the three Spirits. Nor are they aware that they are the main factor in the work. Yes, the Quintessence which they are seeking has already escaped them during putrefaction. Consequently, when they strive with all their might to drive things out with the fire, they also force the Salt with the combustible oil over and the feces as well. Similarly, a portion of the four elements goes over. Thus they spoil one with another, as will be taught later. Neither do they know the elements in their inner and outer nature; nor do they know the fine substance which keeps them together and binds them, with which God has adorned them. Therefore, I am justified in saying that their thing is altogether madness and fraud, and they do in no way understand the work, as has been proven.

CHAPTER XV

Now I will teach my child the formost and principal factor of the Vegetable-work, which is the first beginning of the *Vegetabilia,* since there is nothing nobler nor subtler among all growing things. And among them there is all that of which the *Quinta Essentia* goes over first. Its name is the noble wine. Comprised in it are also wheat and all cereals, all fruit of trees, and everything of which the Qu. Ess. goes over.

I will therefore instruct my child how to make the Vegetable-stone of these, which stone surpasses the mineral and also the animal stone. And it is not *corrosive* like the other two stones. And the gold made of it is not corrosive like the gold that comes from the other two stones. That is the reason why it is the supreme medicine of human life, driving away in a short time all sicknesses that may befall man, of which instruction will be given later on. These are the reasons why it is *secreter,* but none of the other two stones is. It is also easier to produce and requires less time and costs. Therefore, it is *secreter,* while the other two stones are not.

CHAPTER XVI

Let my child therefore take, in the name of the Father, the Son and the Holy Ghost, a good, old, clear wine, of good smell and taste, as it comes from its grapes — not brewed or made but such as has grown of itself, so that nothing foreign (or: alien) may be in it. Neither should there be a mother, druse or yeast in it, but it should be a wine which has been drained three or four times of its druses or *fecibus,* each time into a clean or fresh cask.

When you have this wine, you should have a big kettle built by masonry into a furnace. The latter should be constructed with a snout coming out of one side and extending to the bottom. Into that pipe the water is to be poured whenever it has boiled or steamed away. Into this kettle put a large, earthenware pitcher of 16 or 20 Cologne quarts. Fill those almost completely with wine. Then you should have a broad cork which fits the mouth of the pitcher. Or have a turner turn a large stopper which will just fit into the mouth of the pitcher. Now put a helm on and place a receiver to the spout of the helm. In addition, the *recipient* should have a snout in the stomach, to which snout you must again put a receiver. Into that the noblest spirits will go. If one wishes, one may still put a glass to such a recipient. Then the spirits have room for play and thus much less violence is done to the *lutaments* by the spirits wanting to penetrate out through them.

You could, therefore, as you wish, put four or five recipients, one next to the other, and each time the subtlest spirits will be in the glass. Distill your wine over in such a way, and keep the residual *phlegma*. Pour the spirits or *aqua vitae* together; put them again into the earthenware pitcher; put the cork back into the neck, and put a helm on. Put three or four receivers to the snouts. Now distill gently *per balneum,* and watch carefully for some dew or smoke in the alembic, which is a sign that some wateriness rises together with the spirits; for when the spirits rise alone, the helm is as pure and clear as crystal. But when wateriness rises with the spirits, it shows in the helm. Look carefully, therefore, if you do not notice some hazy vapor in the helm. This will happen during the last distillation, when almost everything has gone over.

As soon as you notice something of such a haze or vapor in the helm, stop the distillation and keep that which remains in the pitcher separately, because you must distill it again. Gather for this purpose all the residue of all the distillations, and distill them again *per balneum* until they rise without smoke. Repeat this till you have all the spirits out of the wine, without finding any steam in the helm. Then you must no longer distill per balneum, but you must still rectify the spirit by fire, in a glass, in a cupel with strained ashes. This must be a long glass without a helm. (The long glass must have a hole above to allow you to pour into it through a funnel, and afterwards it has to be luted).

CHAPTER XVII

Put it, in this way, in the ash together with the receivers attached to it, and distill on a gentle fire, since the heavy spirits sink down on the side of the glass, looking like little veins or streaks. They fall down to the bottom of the glass because they are heavy and coarse while there is still some *phlegma* with them. The subtle spirits move through the spout, while the heaviest part falls down to the bottom of the first recipient. But that which is subtle stays floating in the receiver and wanders through the snout into the second receiver, and so forth into the third, and in each case the most subtle spirit is in the last *receptacul*.

You should also always leave something in the glass which stands in the cupel with the ashes. Make again an infusion of everything there is in the first two receivers; but keep what is in the third alone in a glass, well closed. Distill it again (reducing it) to just a small amount, for a little must remain, and pour it together into a glass. Pour also everything you find in the third recipient together; it is the subtlest, and that which you leave is the coarsest. That must still be distilled often till you have everything together in the third receiver.

If it should happen that you notice some vapor or dew in the helm, you must distill again per balneum, as has been taught before, and afterwards by means of the long glass, leaving each time a little until everything together is in the third receiver, which you must gradually gather in a glass till you have everything of the third receiver together.

Now pour it into the long glass and distill over into a receiver that should have a tube in its stomach. But that tube must be well luted. When it has gone over, put the receiver into the cupel with the ashes. Cover it above with a small piece of cut glass or tile, but open the tube and put another recipient of which also has a snout in the stomach that must be closed tightly. Now distill from one receiver into another, and you do not need to pour (liquid) over it; but, when it has gone over, remove the glass from the ash and put the receiver with the spirits into it. Put the other one (recipient) on again, and in that way distill over and back again.

Or you can have two pelicans made, which are called two brothers. They distill one into the other without an opening. But one has to put them over and back again so that, if one is distilling, the other is the receiver. But it (the liquid) will rise over so easily and in so short a time that you will be surprised, and that will last through 16 or 20 distillations. Afterwards it will gradually begin to become lazier, so that it no longer rises over as fast. Each distillation becomes slower, since the spirits begin to become coarser and thicker, and finally it rises so slowly that the glass will glow at the bottom. It will finally sublimate. It is not necessary, however, to draw it over so long, because it would take too much time. Nevertheless, it is possible, and I have done it myself and have also seen someone else do it, but there are many other ways to reach such sublimation, as I will instruct you later.

Therefore, if it begins to go over slowly and lazily, stop! The wine has then been sufficiently rectified of its phlegma. Put a glass on, stopper it quite firmly and preserve it until I teach you what to do with it.

But now, my child should rectify all the water from which he has drawn his wine, since you must draw your water out of the fire and the earth; as you have drawn it out of the air, so you must also draw it out of the fire and the earth, as will be taught later. That is why Aristotle says: "When you have the water out of the air, and the air out of the fire, and the fire out of the earth, you have the right Art, and from it there comes a stone which is no stone, neither has it the nature of a stone." My child should now take all the water and rectify it per balneum in an earthenware pitcher with a helm. When all the water has gone over, remove the pitches from the balneum and you will find at the bottom a black matter, as black as pitch. In that black matter the element fire together with the element water are hidden, as is the combustible oil together with the dry water, which is *Salarmoniac*. You must take the latter out and rinse the pitcher, taking care that nothing gets lost; and preserve it well. Do not draw the water off dry from it, but leave a little moisture, or a little water, with it. Otherwise you could not remove the aforesaid pitch—black matter clean out of the pitcher. After that you must evaporate it quite dry in another vessel which must be wide enough above, so that you can take it out more easily.

Now pour the distilled water back into the pitcher, and distill it over together per balneum, as before. When it has gone over you will again find a black matter at the bottom of the pitcher. Take that also out clean, let it smoke off, and add it to the previous (black matter). Keep it well. Then distill all the water once again per balneum. You must repeat this till nothing remains. Add what remains each time to the other. When the water goes over pure, without anything left behind, you have the water clear out of the fire and out of the earth.

CHAPTER XVIII

Now put all the black matter into a big, earthenware pitcher of 16 to 20 quarts. Put all the black matter into it. Now set it in the balneum and pour enough of the water you have drawn from it into the pitcher to fill it to one quart. Stir it with a wooden spoon to mix it well. Bring the balneum to the boiling point, but without actually boiling, for one or two hours, till the black matter has disintegrated in the water and become mixed with it. Now cool the balneum to the point where you can put your hand in it. Let it stand at that warmth for two days and two nights, stirring it every three or four hours with the wooden spoon, so that that which is at the bottom rises; always close the pitcher with a fitting and cut little cover.

CHAPTER XIX

Let it not be a secret for my child that then the water will extract the Elemental-fire and become red as blood; neither will it draw anything else into it but the Elemental-fire.

CHAPTER XX

My child should know that from everything God has created and is comprised among the *vegetabiliae,* the air must first be drawn off by distillation per balneum, as has been taught before. When the air has been removed from the water, the water must be drawn from the fire and the earth by distillation, as has been indicated before.

CHAPTER XXI

After that it is no longer necessary to distill with the alembic in order to draw the fire from the earth; for if the water has once been separated from the fire and the earth, it will no longer mix with the fecibus. And even if the feces became mixed with the water, the water will nevertheless push them away and make them fall to the bottom. Instead, it will absorb the Elemental-fire, which is red, and will let the feces drop together with the earth and the combustible fire, in which the salt, or the dry water, or ✳ is locked - the water will let all these sink to the bottom, but it will keep in itself the Elemental-fire, which is red. For if the water has once been perfectly separated from the elements and the fecibus, with which the elements are mixed, the water of the clouds, which is the *phlegma,* will henceforth at no time mix again with fecibus of the world from which it has been separated. That water of which we are now speaking is water of the clouds and not an Elemental—water, as has been proven before. Therefore my child should know that this water draws out everything Elemental from the fecibus of the elements; and it lets the fecibus drop down, keeping in itself that which is Elemental.

CHAPTER XXII

My child should also know that with all things included in *animalis,* be it herbs, spices, animals, cattle or human beings, the water first goes over, and the air and fire are both drawn out simultaneously with the water of the clouds, as will be taught, if it pleases God, in the Animal-work. With everything, however, included in *vegetabilia,* the air goes over first, such as wine, honey, and all cereals, such as wheat, barley, oats, buckwheat, vetch, and some seeds, all fruits of trees, some from herbs, and everything comprised under vegetabilia. With those the air has first to be separated from the water, as has been taught here. Afterwards the water must also be drawn out from the fire, and from the earth, and from the fecibus. Then the distillation has been done perfectly. After that one can draw out from the earth with the water; following that one draws the earth from the fecibus with the water; one also draws the salt or ✳ from the combustible oil, as will be taught later.

Therefore my child should know that in the Vegetable work there must be more distillations, also more labor, more time and greater costs. This does not apply to the Animal work, for in all animal works the water of the clouds goes over first, while the air, fire and earth stay at the bottom with the fecibus. When one has the water pure and clean, so that nothing remains, the water has to be poured on again.

Now the water will draw the fire and the air together from the earth and the fecibus. Consequently, the work takes in everything less time in the animal than in the vegetable work, on account of the reasons mentioned above.

CHAPTER XXIII

Now we will take up our work again. When the pitcher has stood for two days and two nights in the balneum, in accordance with the previous teaching, take it out and let it stand for two or three days in order to sink down. Now have at hand another clean pitcher or vessel. Let it (the liquor) run off its fecibus into it (the vessel) per filtrum or through a little piece of cloth, according to the Art. When everything has been drawn off, take the pitcher with the fecibus and the earth and put it again into the balneum. Again pour its own water upon it till the pitcher is filled up to a quart and stir it (the fecibus and the earth) well into the water with a wooden spoon, as before. Then remove it and let it sink again for two or three days, so that the feces and the earth settle down at the bottom. Now separate it again *per filtrum* and pour it to the first. Cover it, then put the pitcher with the fecibus and the earth back into the balneum. Again pour some of its own water upon it, and proceed in everything as you have been instructed before. You must repeat this drawing off, sinking and filtering till the water is no longer colored but stands pure and clear above the fecibus. Then you have the fire from the earth. Take the feces mixed with the earth from the pitcher and put them into another vessel.

CHAPTER XXIV

Now take all the water in which the fire is and put it into a big kettle. Take a large quantity of egg white and beat it as thin as water. Take some of the water in which the Elemental-fire is and stir it for a while with a (wooden) spoon in a pot or pitcher, together with the egg white. Now pour this mixture to the other water in the kettle, stirring constantly, so that the egg white is well mixed with the water and the Elemental—fire that is in the kettle. Now put it over the fire and let it come to a boil on equal (steady) heat. The egg white will coagulate; and should there be feces left in it that did not sink down, the coagulated egg white will attract and purify them. Skim and drop them; then draw them off *per filtrum* or hang them in a "claret-bag" and let them separate (drip off) well. Now you have your water and your fire pure and clear.

CHAPTER XXV

Aside from this, there is still another way for clarifying. When you a*bstract* the water per balneum, the fire stays at the bottom, while the fire in the balneum does not rise. Then pour the water again upon it and mix them well, and it will drop its feces. Now draw it off again per filtrum and distill the water from the balneum, as before. This must be repeated so often till there are no more feces. Then you have both your water and your fire pure, and this is the best way, but it takes longer and costs more on account of the fire. The feces are each time added to the first fecibus from which the fire has been drawn. In the same way, the feces with the egg white are added to the first. My child should know that if the water has been drawn off the fire, the earth and the fecibus, and is poured on again, it does not absorb anything except what is pure; and it lets the impure sink. That not only occurs in this work but in all works where the following is done:

The water, which was thus rectified beforehand, so that it does not leave any feces, is drawn off;
The same water is poured on again, be it on fire, earth, salt or ✳, any of those well mixed with the water so that it dissolves;

It stands for one or two days to let the feces sink, since no feces dissolve in it;

The water is poured back again, as has been taught before.

Now the fire or the earth or the ✳ can be brought to crystal clearness, coagulated hard, the one harder than the other, since the fire may well be coagulated in clearness but not in hardness. It becomes like cheese which may be bent when it is not old; and if one takes a piece of it thick like a finger, one can well stretch it. Or like a piece of horn from cows or oxen, which has lain for some time in boiling hot water — that may also be bent. Thus it is also with this element of fire. It becomes dry, hard and clear, like crystal, red like a ruby, and yet it is not brittle but can be bent. That is due to the humidity of the elements water and air which are in it and mixed with it, as was proven before, so that air and fire should not be separated. But when the element earth has been clarified with the water of the clouds, it is hard, dry and brittle, clear, transparent and white like crystal, because the element earth is cold and dry.

CHAPTER XXVI

When the ✳ has been clarified with the cloud-water, it is likewise white, clear and transparent, hard and brittle, because it is hot and dry. Therefore my child should know that one can bring everything in the world to crystal clearness by the water of the clouds, once it has been drawn off pure

135

from that which is to be brought to clearness, but it must not leave behind any feces. Then (if it should still leave feces) it is poured on again and well stirred. After that, it should be allowed to settle down, and then it will discard its feces, keeping within itself that which is *perfect*. Now it has to be poured off the fecibus per filtrum and distilled over *in balneo per alembicum* until it is dry. This work has to be repeated so often till no more feces are left behind. Then, dried again, it becomes hard, clear and transparent like crystal, as has already been taught and will again be taught hereafter (if it pleases God!)

This is the right way to bring all things to crystal clearness, not only in this work of the wine, with which we are dealing here, but in everything God has created, in human beings, cattle, birds, fish, animals, herbs, flowers, fruits, metals, stones, and everything that exists under Animal, Vegetable, and Mineral. Among those three all things are comprised that God has created in the world. And thus it is possible to bring everything created to crystal clearness by means of the Art which God gives to his children and lovers. For after Judgment Day, God will separate all things and make them clear like crystal and red like rubies. After that, no corruption will enter them again, and they will last in all eternity.

Do you believe, my child, that everything created below here by God will pass away at Judgment Day? No, not the meanest little hair God ever created will go under, no more than the incorruptible heaven; but God will transform everything and make it crystalline according to his will and pleasure. Therein the four elements will be perfect, simple, fixed and unchangeable, and then everything together will be *Quintessentia* and *Lapis philosophorum.* That can be proven here in this world by the Art, by our ability to bring crystalline clearness to everything created by God, and by reuniting the four elements into a simple and fixed nature, so that they can afterwards not be changed by anyone. Neither can they be transformed or burnt by fire, but they will remain in all eternity as they are. And all this may be accomplished by human intelligence and subtle mastery, God having granted his children such wisdom through special Grace.

From here it comes that the hordes of philosophers say in their books that the Art is in everything God has created, by which they speak the truth; but they withhold information on how one is to draw it out, and the ignorant can therefore not understand their words because of the darkness of their intelligence. The fact, however, that I am quoting those discourses so extensively is for the purpose of letting my child understand all things at bottom and know what he does. Also, if you should make a mistake in your work or if it went wrong, or if you had missed something, or one or another defect had occurred in your current works, you should know to what it is due and where you went wrong, so that you can easily correct the matter. Therefore, my child, read and reread this book often and understand it well at bottom. Here nothing is presented to you in parables, or communicated to you in some dark words, but it is told you in its proper meaning, as you yourself can test with your own intelligence, lest you should fall in error.

CHAPTER XXVII

Now we will return to our work. After rectifying the fire either with egg white or per balneum, so that no feces remain, put it finally into a large Venetian glass. Draw the water off clean till it is completely dry; then let it cool down. After that, you must break the glass. Take it out, and your fire is as clear as a crystal, red like a ruby. Keep it till you need it.

CHAPTER XXVIII

Now have a potter make a large earthenware vessel
for you of good earth. It must be fireproof. Have it
well glazed with lead (verbleyglasen) on the
outside. On top of that, lute with a good lute on
the outside, about two or three fingers width. Let
the lute dry well. That vessel should be made thus
(see picture).

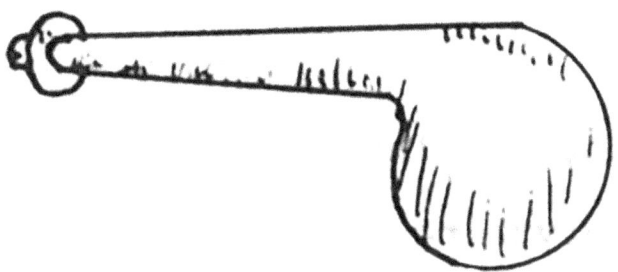

Put in it all the feces you have in which the earth
is still, and the combustible oil plus the ✳. In
addition, you should have a large stone pitcher of
20 Cologne quarts. Fill this pitcher half with your
cloudwater.

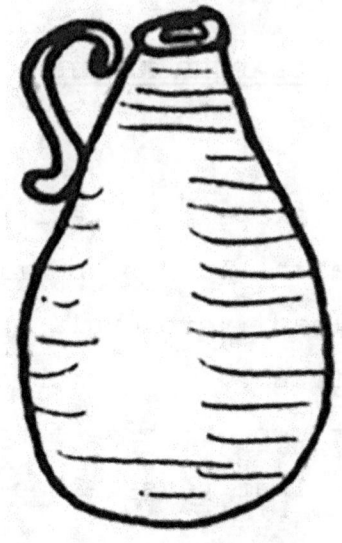

Have ready a suitable furnace. Put the
aforementioned vessel in it upon a grill, so that
the flame can reach it all around. Stuff the mouth
or the neck of the vessel with hay and tie a cloth
around the mouth of the vessel to prevent the hay
from falling out. Put the pitcher containing the
water on the neck of the vessel to prevent the air
from escaping. Let the lute dry well before you
light a fire in the furnace. After the lute has
completely dried, light the furnace, first with a
gentle fire, for six hours, so that it (the matter)
gets warmed thoroughly. Then increase your fire a
little for another six hours, so that your vessel
with the materia be heated through and through. Now
increase your fire considerably, so that your vessel
begins to glow, meaning that your vessel is heated
through after six hours. Subsequently, increase your
fire so much that your vessel begins to glow

strongly for five or six hours. Then let it cool
down of its own. Now remove the can from the mouth
of the vessel and you have in it the combustible oil
and the salt, or ✳, which lay hidden in the
innermost of the combustible oil and was mixed with
it as also with the earth and the fecibus, from
which they have now been separated by the great heat
of the fire. The fire has driven the combustible oil
and the ✳ from the earth and the fecibus, and the
combustible oil is swimming upon the water, black
and thick like lees. The ✳, however, which has now
been separated from the combustible oil, has blended
with the water and made it white like milk; and it
is very corrosive on the tongue.

After that, you must take a large, glazed cupel,
burnt of clay. Into it you must pour everything
there is in the pitcher. Make your water boiling
hot; pour some of it into the pitcher and rinse it
till it is clean, because the combustible oil clings
to the sides of the pitcher. Pour everything
together into the cupel. You must rinse the pitcher
so long till it is clean, and then pour everything
together into the cupel.

CHAPTER XXIX

Now you must have a vessel made of wood, but it would be better if you had one made of earth by a potter. Let it first be burnt, unglazed; and when it is burnt, let it be glazed with two parts of minium (red lead), one part of copper ashes or copper slag (Schlag), and ½ part of tin ashes, rubbed together with salt and ashes. No corrosives can penetrate this glazing, so tight is it. Have all your other earthenware pots that you require for this work glazed in this way and burnt. Such a vessel should be made in the following manner: The lid should be made of wood or stone; the body (or: stock) with the plate in which there are the holes, should be made of wood.

Now skim the combustible oil neatly from the top and pour it into this vessel. Pour the water back into the pitcher from which you had first poured it, and keep it till I teach you how to rectify the Salmiac (ammonia). Remove also the earth from the long retort and keep it till I instruct you as to what you should do with it and how you are to calcinate it.

When the combustible oil is in this cask, take some of your water and make it boiling hot. Pour it boiling hot into the cask upon the combustible oil, and quickly put the body (stock) with the plate (or: disk) and the holes into it. Put the lid on so that the stock enters the hole of the lid. Close it up around and around with a linen cloth. Start pumping, and push up and down like women churning butter. Do that for a good half hour. Then stop, remove the lid, take out the stock. With hot water wash your stock and lid clean of the combustible oil adhering to them. Whatever you wash off, add to the cask and let it settle down for one day and one night.

Now take a large, well glazed earthenware cupel and draw all the water off through the tap, until the oil begins to come. Then stop. Pour the drawn off water into the pitcher in which the water with the ✳ is, since some ✳ is still with it. After this, you need no longer take of your water, but take only common distilled water. Make it again boiling hot and pour it into the cask upon your combustible oil. Again start pumping, or churning, for a half hour. Then stop. Rinse your lid and stock with the water and pour it into the cask. Let it again settle down for one day and one night, and then draw the water off into the cupel till the oil comes.

Should it happen that some oil were running out of the tap together with the water, remove it neatly and put it back into the cask to the other oil. Pour the water into a pitcher or vessel by itself, for there is still something of the element earth in it, which has gone over with the combustible oil.

When the oil is thus clear, all the water must be evaporated in order to calcinate the earth contained in it along with the rest. When the ✳ is rectified, the earth coming out of it should also be added to the other, so as to calcinate them together. Then take again boiling hot, distilled water and pour it into the cask, and pump as before. Draw it off, and put all the water into a cask; again pour other hot, distilled water upon the oil, etc. Do this till the water runs off as clean as when you poured it on. Then the combustible oil is well clarified. As to the water which you have all poured together, evaporate it; that which remains, add to the earth in order to calcinate it along (together with the earth).

CHAPTER XXX

This combustible oil which you have just rectified, is now clear, thin and red like blood, also greasy like other oils, and is hot and humid. It is used to anoint or rub nerves in which one has a cold or stiffness (or: gout, arthritis); also lame, chilled members (arms and legs in which one has a cold or rheumatism). Likewise, it is good for persons who have the "drip" (Tropfen)[2] or a stroke (Schlag). It is also used in all ointments and poultices for *incarnating* or causing flesh to grow in all deep holes and wounds.

CHAPTER XXXI

It serves my child to know that, if this combustible oil did not exist in all the things growing out of the earth, we could not live, just as we could not live without the water of the clouds or the rivers, since without water no food can be prepared and no medicine blended with another. Neither could dyers put color in cloth, if it were not done by means of water. Whatever one wishes to do or prepare in the world, water must always be there. If there were no water of the clouds, people would be helpless. But we could dispense even much less with the combustible oil. For if there were no combustible oil, nothing in the world could grow from the earth, neither cereals nor fruit, neither trees nor herbs. Nothing in the whole world could grow, since the mother of the combustible oil is the fattiness of

---

[2] Probably "dropsie." -HWN

the earth from which all fruits take their nourishment. For if the combustible oil were not in abundance in the earth, corn and all cereals, seeds, trees and herbs would have no combustible oil in themselves. Now then, however, corn, cereals and everything growing out of the earth gradually draw the combustible oil from the earth, each as much as it requires, until it has reached its full growth. After that, it no longer *attracts* but starts withering. For example: Sow corn or the seeds of other herbs (wither) into nothing but sand in which there is no combustible oil. Nothing will grow or green from it but it will dry and come to naught. See what happens when the farmer plants his field for six or seven years in a row without putting manure in it. Such a field becomes arid and meager that fruits finally no longer grow in it, just as is the case in sand. This is due to the fact that the fruits which it had, had attracted all the combustible oil, and that finally there was none left in it and nothing could grow in it. If instead, there is a piece of land on which grass is growing and it is left without being attended to and it is not grazed bare, letting the grass rot on it through the winter so that the combustible oil thereby seeps back into the earth, then watch how luxuriant and fat the land becomes when the combustible oil doubles from year to year.

Consequently, we cannot live without the combustible oil, since we must take our nourishment from it. Nevertheless, it is also the cause of our death. Observe if the revelers and gluttons live very long, for they take into themselves more than is necessary for their nature. And as we take more food into us than our nature requires, bad and harm-full *humors* arise within us, such as blood boils (Bluteissen), abcesses, cancer and fistulas, or other bad ulcers, and many kinds of sicknesses whose principal cause is that there has accumulated too much combustible oil within us, for the reason that our nature attracts too great a quantity of it from our food and drink than nature requires. For example: if a Master or surgeon puts an excess of fatty oil into a wound he wishes to heal, a rank growth of flesh starts under his bandages and bad flesh will then grow there If he persists with the same ointment for a long time, not reducing its fattiness, *corruption* and *putrefaction* will arise in the nerves and flesh, so that finally, fistulas, cancers, and running holes (sores) will well up in it. All of that is caused by the combustible oil, for all oils and fattiness take their origin in this combustible oil which they have drawn from the earth, and thus it follows as a consequence, that the combustible oil can also be the cause of our death.

CHAPTER XXXII

Combustible oil is also found in mines. It is called *Sulphur,* because the philosophers call *Sulphur* every combustible oil found in the elements; and they say:

Our ♀ is not common ♀. That is, our Sulphur is incombustible - whereby they are speaking the truth; for when they say that our Sulphur is incombustible, they mean the Elemental—Fire which is extracted from the combustible oil. <u>That</u> is the Sulphur they mean. Thus there is in all things in the world combustible oil that is not perfected (perficiret). Yes, in all metals there is combustible oil, except solely in gold in which there is no combustible oil. Because of this, it is also fixed and, therefore, the fire cannot destroy it; about which, sufficient has been said in the material on the Mineral—Stone. I am speaking so much about the combustible oil, my child, so that you should understand the nature of all things and know the elements and what is mixed with the elements, so that you may know the inner and the outer and thus, not make errors in the work you undertake. And if by chance you should make errors in your work, or if you had neglected certain matters, you should then know what kind of a mistake it is and how you can correct it. Therefore, understand all my words and their meaning well, so that you will not go astray.

CHAPTER XXXIII

Now we will resume our work again and thus undertake to rectify, or clarify of its fecibus, the Salt, or dry water, or our ✳ . Therefore, take the pitcher containing the water with the ✳, from which you have skimmed the combustible oil. Put it in the Balneum with a helm attached, and distill all the water off until the matter is dry. Thereupon, remove the helm, pour the water back upon the matter, and put a cut piece of slate (or: shale) on the mouth of the pitcher. Let it stand for two or three hours in the Balneum; take it out and let it settle down (clarify) for one day and one night. Following this, filter the clarified water from it and pour this clear water again on the feces. Stir it and allow it to settle again. Once more, filter the pure (clear) water from it and add it to the first water. Now test your feces on your tongue to see if they are still sharp. If you still find some sharpness, you can pour some more of your water upon them and proceed as before. When no more sharpness is discerned, add your feces to the earth in order to calcinate also what had been driven over by the strong heat of the fire at the time that the combustible oil came over together with the *Salmiac*. Now take all the water and pour it once more into the pitcher, then, put it into the Balneum with a helm attached, and draw the water off until it is dry. Remove the helm, pour the water back on it, and let it stand for three or four hours in the Balneum; which should be quite hot, so that the Salmiac can be well dissolved. Now remove the pitcher, let it stand for one or two days in order to let it settle, and again filter it of its fecibus. Do this until no more feces are left, then it is sufficient. Finally, draw the water off through the helm in the Balneum, till the ✳ is dry. Each time, add the remaining feces to the earth in order to calcinate them along.

Then, when your ✳ has thus been drawn off dry, take it out. It is as white as snow. Put it into a glass bowl and set it into your dry living-room. For if you left it standing in cold air, it would dissolve[3]. Put it near your Spiritus, or Air, and near your Elemental—Fire, and preserve it well until I teach you what to do with it.

CHAPTER XXXIV

---

[3] "per deliquim" -HWN

Let my child be informed that just this ⁂ is the Salt of the Wise, of which the multitude of the philosophers speak about so often and so covertly in their books. Without this Salt, no Philosopher's Stone can be prepared, for if this salt were not a part of the Stone, it would have no *ingress*. Very often they refer to this as our *"dry water"*, for without water, there can be no <u>composition</u> in the world, to bring one thing into another, as been mentioned previously. Thus it is called by them, their "dry water" and when simpletons read this in

the books of the philosophers, they think it is ☿. This leads them into a great error. Therefore, the Sages have given many names to the Salt, so as to hide or obscure it. They also call it the *Salt of the Wise,* sometimes also, the *Flying Eagle.* Then the simpletons believe that they have understood the

Salt to be ☿, and think they have comprehended the words of the philosophers, yet they are sorely mistaken.

CHAPTER XXXV

Now we will again return to our work, which is, to *calcinate,* or to *reverberate,* the earth. To do this, take your earth, put it into a flat, earthenware

first burnt in wood from fern and other herbs, wood or the like, straight into the furnace, using such great heat that it had to melt, it would turn into a black, ugly, dark and opaque glass. For the element earth has curdled (coagulated) together with the fecibus, and although it were standing in the fire for a whole year, it would not become white, because it is a firm *compact-corpus;* thus the earth coagulated the feces.

That is why the glassblowers must first reverberate their ashes till their feces become white as snow before heating their ashes so much that they flow; for as long as the ash does not come into flux, the feces with the corpus of the earth are open, so that the heat may well burn through till the feces are white as snow, for then it is easy to make white, transparent glass of it.

Why am I telling this to my child? So that you should know that the element earth cannot be burnt by fire, for it is an Elemental-element. If it had no feces in it, and if it were heated till it would melt, and even if a cupel were as thick as ten shoes, and there were enough earth, it would penetrate through it. Therefore, my child, if you failed in your work with the fire, you should know how to get your Elemental-earth back out of the fecibus.

CHAPTER XXXVI

Now we will again return to our work. Give heat, therefore, and let it stand in heat until your earth is as white as snow, which you can see in the following way: Lift a pan out of the furnace with tongs and let it cool down. Then you will see if it is white as snow. If it is not yet that white, put it back again till it is white. Then remove it and put all the earth of the pan into a large Hessian or Venetian glass, but not into a stone pitcher; for when the earth is dissolved in its water, it would penetrate through it (the pitcher), even if the pitcher were as thick as ten shoes. So subtle is the element earth when it is freed from its fecibus. It is also the smallest of all elements, yet the most subtle, of which more will be taught. Therefore, put it into a large, double glass, pour your water upon it, and put it in the Balneum for one day and one night. Let the Balneum boil, then cool down. Take it out, put it aside, and allow the feces to settle during two days and two nights. Now pour everything carefully down into another large glass, by bending the first to one side (decant). Again pour some of your water over the feces and put it back in the

Balneum as before. Proceed in everything as before, and again pour it off into the glass by *inclination* (of the first glass) to the previous. For the third time, pour water upon the feces and do everything as before; then remove the feces. Set the glass with the earth in the Balneum with a helm, distill the water off till the earth is as dry, that it is like dust. Now let the Balneum cool down, pour your water on again and give fire for three or four hours till your earth dissolves. Then remove your glass again from the Balneum and let it settle for one day and one night. Now pour it off again per *inclination* from its feces into another glass, and put it back into the Balneum with a helm; again distill the water off till it is dry, as before, then pour it back on again and let it dissolve as previously. Remove it and let it settle as before, and again pour it off by inclination (decanting) from its fecibus, as before. Repeat this work of pouring on and drawing off till no feces or residue remains. Then finally *abstract* so dry that it becomes like dust, then you have your Elemental—earth pure and clear, and as white as snow.

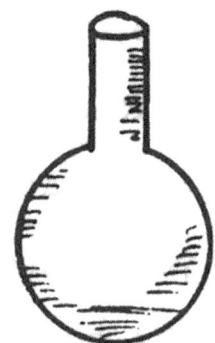

Now my child should know that one may well *clarify,*

or *rectify,* this earth and the , as also the
Elemental-fire, with egg white, as has been taught
before, but it must in so doing be purified about
three or four times till nothing is left of the
feces. You should test it in the following way:

Take a small glass each time and evaporate the
water. Then pour other water on it and let it
dissolve. If it does not leave any feces, it is
enough; otherwise you must purify it better. It
would also be necessary, after you have purified it,
that you should pour water on the feces and the egg
white which you have skimmed, or which stayed at the
claret-bag, if something elemental were still in it,

either of the fire, the air, the or of the earth
which you have purified. That is why it would be
good if you were to pour some of your water and let
it extract and settle, then pour it off again by
inclination or per filtrum, and add it to that which

you are purifying - or you can once again purify it by itself before adding it to that which has been purified. This purifying is the worst way, for one may well carry out this purification twice a day; but clarifying, as has been taught before, is best and most useful though it takes longer. Therefore, you may choose what you like best.

CHAPTER XXXVII

Now we will again turn to our work and steep the spiritus or air in its earth and its ✳.

Accordingly, let my child take the earth and the ✳ and rub them quite dry on a stone. This must be done in a dry room, so that no cold or humid air, nor watery humidity, will be added to it. When one has been blended with the other in this way, you should put it into a glass hanging lamp, or into a glass of the shape of an egg, as is shown in the picture. Pour the spirit over it and let the spirit or air

imbibe into its ✳ and earth, till everything has been absorbed. Then seal the glass with *Sigillus Hermetis* and hang it into the secret furnace. Give it fire of such heat that you can keep your hand in it, without injury, (by sticking it through the hole in the side of the furnace), for the duration of one Ave Maria. Let it stand in such heat for twenty days and nights. Then let it cool down, remove it and break the glass. Now the air or the spirit has congealed with its ✳ into a hard, clear, transparent stone, white like crystal, because the element-fire is not yet in it. Take it out, crush it to a subtle powder, put it into a glass pot with a strong bottom which must be wide below. Add a big helm and a receiver to its snout, and lute it quite tightly. Start a fire in the furnace, gentle to begin with, and gradually stronger by degrees, till the spiritus and the ✳ sublimate together in the form of a clear crystal, and white as snow. When everything is sublimated, let it cool down, remove the helm and break it into pieces, because the sublimate is clinging so much to it that the glass has to be broken or else it (the sublimate) cannot be taken out. It is due to the ✳ that the spiritus must attach itself so firmly together with it. But if one were to sublimate the spiritus alone through its earth, without adding the ✳, the spiritus would not settle on the glass but would sublimate like snow. Now take it out and keep it in a dry room. Remove your earth from the glass pot and dissolve it in your rectified water. Small white feces will settle at the bottom. Filter the water off them into another glass; set that in the Balneum

159

with a helm, and distill all the water off to a dusting dryness. Now take it out and rub it again in your dry room with the spirit and the ✳ which have been once sublimated together. Put them again to sublimate, as has been taught before. You must repeat the sublimation so often and in the same manner as you have already been instructed, till your earth does not leave any more feces. Then your spiritus, your ✳, and your earth are well rectified and ready for your work of making your stone.

But my child might wish to ask: Was the spiritus or air with the ✳ not well rectified before you blended them with the earth? Was the earth not well rectified before you made a conjunction of all three? In reply, it will serve my child to know that they may all three have been well rectified before they were commixed; only, it shall not be concealed from you that there are two kinds of feces in all things created here below by God, one exterior (kind) and the other inside in the depth, which cannot be brought out unless the exterior feces have first been discarded. After that the thing whose inner feces one wishes to extract must first be calcinated; and when it is calcinated, its feces can also be drawn out of its deepest or innermost. As long as a thing has not been prepared in this way, just as long it is not suitable for making the stone of it, neither in the Vegetable, the Animal, or the Mineral.

The calcination, however, takes place in the secret furnace or *tripod*. There the spiritus or air is calcinated together with the ✳. Then, during sublimation, they leave the feces which they contained in their innermost. Now put them into your dry room and keep them well, till I teach you what to do with them.

CHAPTER XXXVIII

Now we will return to our work. Take the Elemental—fire and put it into a glass. Pour some of your rectified water upon it and let it dissolve in the Balneum. Then put your earth into another glass and also pour some of your rectified water upon it; let it likewise dissolve in the Balneum. Now pour the two waters together, mix them well, put them in the Balneum and distill the water off to dusting dryness. Then remove them and put them into a glass hanging lamp or egg, as you did with the spiritus and the ✳, and seal the glass with Sigillus

Hermetis. Hang it into the secret furnace for twenty days in order to be calcinated. Give it the same heat that I told you in regard to the calcination of the spiritus and the ✳, or a little hotter, because fire is not as volatile as the spiritus and the ✳. Therefore, you can give them so much heat that you could hold your hand between the walls of the furnace and the vessel. After it has stood for twenty days, take it out and break the glass. You will find the earth and the fire hard, red and clear like a ruby.

Now put them into a glass and pour some of your rectified water upon them. Dissolve both in the Balneum; then take them out and let (the matter) settle for two days and two nights. Filter the water off; again pour some of your rectified water upon the feces and stir well. Let it settle again, filter and add to the previous, removing the feces. Now abstract the water per alembicum to the point of dryness. Now again pour the water on it and dissolve it as before, and filter again as before. Repeat this work till no more feces remain. Then again draw the water off as dry as you can; take it out, and you have blended your earth in such a way that it can never again be separated. You also have the spiritus or air and the ✳ together, which can likewise never again be separated from one another, and all are rectified of their outer and inner fecibus, and prepared to make the Vegetable-stone of them.

But my child might ask: Why do you not calcinate the Elemental-fire simultaneously with the spiritus and the ✳? You should know that there are two elements which are fixed, and there are also two which are volatile, and yet the elements are so mixed together that one cannot well separate one from the other, as has been proven before. You should also know that air is warm and moist and has a lot of water in it and not much fire. Air and water, however, are both volatile and are spirits. In contrast, fire is hot and dry and has not much water in it. Nevertheless, it also has water in it, because there is also air in the fire. One does not find air, however, but water. Consequently, it is mixed with the fire, air and water; but there is not so much of it that air and water have the power to raise the fire out of the earth during sublimation. When therefore, fire is with earth, which is also fixed, the element earth retains the fire so that it cannot be sublimated, for earth does not *participate* either with the air or with water, as fire does, since one

may well separate the air and the water from the
earth, as much as possible.

Nevertheless, my child should know that there is
also earth in the air, in water and in fire; for if
there were no earth at all in them and if they were
pure spirits, they would be invisible and
intangible, and they could not be coagulated,
grasped or seen (or: touched). But they do not
contain so much earth that it would hinder them when
rising during distillation or sublimation. It is not
so with fire, however, but fire contains a great
deal of earth, because both, that is earth and fire,
are fixed. But it does not have as much air and
water in it that would rise, for it would have to be
with the earth when the fire is calcinated with the
air and the ✳. One cannot calcinate elements
unless they are mixed with the earth, otherwise they
would volatilize; and if they were calcinated, the
fire with the air and the dry water or salt could
not fly up. In addition, these two volatile ones
would not drop their innermost feces as one tried to
draw those out of them through dissolution and
coagulation, as one did for fire, because they
contain a great deal of humidity. That is why the
innermost feces of the two volatile parts must be
drawn out by hot dryness, on account of their
humidity; otherwise they would not allow their feces

to draw away from them. Fire, too, would not wish to let go in hot dryness, the feces which are contained in its innermost, because it itself is hot and dry: Its body would only close up during sublimation. Yes, if one could sublimate it, which one cannot do, it would only more strongly hold on to its feces, for one must open up a contrary nature with another contrary nature. How would you open up a hard, *compact* thing with another compact thing? Such is impossible. Rather, all things must be opened up by their *contrariis* (opposites), and be brought out of their nature by another nature which is opposite *é diametro* to their nature. How would you fix a volatile thing by another volatile thing? Such is impossible to do, which fools cannot understand, from which arises their great mistake.

Therefore, my child, heed these words so that you do not go astray; follow nature as much as you can, so that you make no mistakes. Fools sometimes believe that they are opening a thing and yet may close it tighter than it was before, because they do not follow nature, and thus they lose everything they spend (on their work). That is why, in accordance with the reasons quoted above, fire is not calcinated with the spirit or air. Let my child take all this well to heart and reflect upon it often.

CHAPTER XXXIX

Now we will again resume our work in order to bring
these elements which we have rectified completely to
their highest and utmost rectification. You should
therefore take a large recipient, which should be
very long. Into that you must put the air which is

blended with the ✳ and forms one Corpus with it.
Pour some of your rectified water upon it, a little
at a time, and set it in a furnace, in a cupel with
strained ashes. But before the mouth of the
recipient must be cut with a level stone. Then one
has to form a small glass according to the large
glass, with an iron instrument, as the glassmakers
do. One can also grind such a glass on an even
stone, subsequently put it on the mouth of the large
glass, and a leaden weight on top of it. After that,
start a fire in the furnace, first a gentle fire
till your materia gets warm. Let it stand thus in
warmth for twelve hours and it will dissolve as if
it were a red ruby. Should not everything dissolve
in the water, heat some of your rectified water to
the same degrees as your materia in the glass, and
pour it into it (the glass) by a glass funnel. Let
it stand another twelve hours in the same heat as
previously in order to dissolve. If not everything
is dissolved, pour some more of your warm rectified

water upon it, and continue this until all your
materia is dissolved into a clear, red water.

Now my child would like to ask the question: Why did
you not pour the first time enough rectified water
upon it that it could dissolve (the matter)? You
have to be instructed, however, that no more
rectified water can be poured upon it than is
sufficient to dissolve correctly; no more, for there
must be no more moisture of the cloud—water in it
than to allow it to dissolve rightly. That is
enough, for it does not require more cloud—water or
moisture. If one is to merge one thing with another,
it must be done with cloud-water; and then when it
is dissolved, what more moisture does it need? If
then they are to blend in order to stay together
eternally, so that they can never be separated, the
dry water must be there too; that is their Salt of
✳. Otherwise they will not stay together and be
loath to grasp each other thoroughly, even if all
elements have already been well rectified. But if
they did not have with them the dry water, the
elements would not wish to fix each other, and if
one were to fix unto God's Judgment, one could not
fix any Spiritus or *Corpora* with each other without
the dry water which is the ✳; for the dry water
causes the spirits and *corpora* to merge with each

other, and dissolve, one into the other invisibly, just as the cloud-water dissolves two things which are opposites in order to blend them together. The same thing is done by this dry water in an even, invisible way; although we do not see this with our eyes, this dry water nevertheless dissolves the spiritus and corpora thoroughly, so that they never again separate. If the ignorant understood the secret of the *materiae* and knew this dry water, which is a mediator between the spirits and the bodies, all their work would be crowned with success.

Therefore, my child, there have to be two solutions, one from outside, or from the water of the clouds, which one can see with one's eyes; the other from inside, with dry water, which is invisible, if a right solution is to take place.

My child would like to ask in addition: You are teaching that one should not give more cloud-water to a thing one wishes to dissolve, than is necessary just to dissolve it, and not more, why that? And if one were to pour in more water, would then everything be spoiled? In reply, you should know that then it would not be spoiled; yet if you take more cloud-water than necessary, you must draw it off again in the Balneum, for afterwards one has to dissolve ☉ in it. If then there were too much cloud—water, more than necessary, no harm would be done. Consequently, one cannot spoil it with it, for if there is too much of it, one can again draw it off in the Balneum; therefore, it does not spoil.

My child might now also say: You have told me about the cloud—water in order to carry out the external solution, and you say that no blending of the bodies and spirits can be accomplished except by means of the dry water which dissolves the bodies thoroughly and quite invisibly. Should it now happen that one wished to join a body and a spirit so that they should stay together, and we had no dry water, how much dry water would one have to add to bring about such a union? You should know that if you wish to blend a spirit and a body, but do not wish to separate the elements, you must put the spirit and the body together in water, each by itself and each in a special glass. Then you must take half as much dry water and dissolve it evenly in clear water that has no sediment. After that, you must pour all three waters into a glass, stirring them well together and then let them stand on warm ashes, so that they dissolve all three together in pure water without feces at the bottom. Then they are well blended, of which sufficient instruction will be given in the Mineral—Stone.

You should also be informed that, should there be too much dry water in a thing for blending them by dissolution, it does not matter. For if they became fixed, they would not retain more than necessary and what they could fix in themselves. They would let the rest fly away. This is said about the manner of blending a spirit and a body when the elements are not separated. But where one separates the elements, be it in any of the three stones, no dry water must be added, for there is already dry water in it, since in all things in the world there is dry water; it is their salt, as has been taught before. Therefore, understand all my words thoroughly, so that you do not make mistakes.

CHAPTER XL

Now we will again continue our work. When now everything is dissolved in pure water, the water will be red, clear and transparent. Let such water stand on the furnace in gentle heat, just as warm as the sun shines in March, for three weeks, to blend the elements well, one with the other. Then, after three weeks have passed, they will be blended so much that one cannot be separated from the other, and yet they will not be fixed but be between fixed and unfixed. Nevertheless, they will not rise in the Balneum; even if they stood in it for a whole year, continually boiling, nothing of them would rise. But if they were put into a strong, thick glass into a cupel with ashes, heating them so strongly that the glass pot would start to glow at the bottom, and were left thus in even heat, it would gradually rise in the form of a red, transparent oil, clear like a crystal, also red and transparent; and as soon as it got cold, it would coagulate into a red stone, clear and red like a ruby; and it would last in the air, but disintegrate in heat and dryness. And that is how it should be. Therefore, after it has stood for three weeks in the ashes on the fire, you should pour water from the big glass into a glass bowl, set that on warm ashes. Let the cloud—water steam off, and a dark yellow powder will remain, reddish. Now take a glass retort, put your powder in it and put the retort on a furnace in a cupel with strained ashes. Have at hand a glass recipient; attach it to the retort and lute the joints tightly. Now start a fire in the furnace, at first a small, gentle fire; increase that fire by degrees until the retort starts to become red with heat. Let it stand in such heat till everything has gone over into the receiver. Now remove the receiver and put it into a basin with ashes. Heat it, it will melt like wax. Then take a small glass and pour it into it as long as it is warm, and quickly when it is cooling down. Then it will be clear, as clear as a crystal, red like a ruby, and transparent, also half—fixed and half—volatile. It will curdle in cold air and flow in fire. Thus then the Vegetable-Stone has been done. Thank God for his wonderful gifts which he has bestowed upon his philosophers.

This then is the stone which cures all sicknesses which may come into man's body, miraculously in a short time. If you give every day, one grain of it with wine, you will see more miracles than you can believe. Plenty enough is said to intelligent people. But at the end of the three stones, when their projection will be taught, more details will be given on its possibilities. Thus, my child, have I now taught you to prepare the Vegetable—stone, which is the foremost among all three stones.

CHAPTER XLI

Now I will also instruct you in increasing its power a thousandfold. To this end you should take it and dissolve it in your rectified water in a glass vessel, and coagulate it again in the following way: Powder the stone, or break it into small pieces. Put it into a glass pot with a wide mouth. Grind the mouth even on a stone; also grind a small, round piece of glass like it, which is put on the mouth. Put the stone into it and put as much of your rectified water on it that the stone is almost under water. Set it on warm ashes, and it will immediately dissolve. Now put the small piece of glass on the mouth of the glass pot, and let it stand thus dissolved for twelve hours. Now remove the lid, increase your fire and evaporate the water till the stone is dry. Then dissolve it again with your rectified water, as taught before, and let it stand thus dissolved for another twelve hours. After this, congeal it as before, and repeat this work till nothing will congeal but remains as an oil. Then it is ready for dissolving ☉ in it.

Take fine gold, and *cement* it three times in *cemento reguli.* If my child were now to ask: Why should gold be cemented, seeing that it is fine? You should know that something must be added to the gold from which one wishes to get money or coins, or else the coins would be far too soft and too flabby in the hands. That is why one has to cement three or four times to be more certain that it is fine. Further my child might ask: How and why is it that something hard becomes soft through being often dissolved and congealed, getting the consistency of an oil, and no longer curdles, as you have here taught? My child should know that one cannot turn anything in the world into oil as long as it contains any feces, either outside or inside. But when it has been freed from its fecibus, one can turn it into oil by dissolving and congealing it often. For by often turning into water and becoming disembodied, each time being brought back into a corpus, it will become so subtle and volatile that it disembodies by itself. Finally it becomes so subtle that it cannot be retained in any glass. In time, on account of its great subtlety, it would penetrate through the glass as oil penetrates through leather, no matter how thick and hard the bottom of the glass would be. This is why something hard may well be changed into an oil, because of the reasons given.

CHAPTER XLII

Now we will again revert to our work. Take the thus cemented ☉, as has been indicated; have it beaten into thin leaves, as painters need for gilding. Rub those leaves on a marble with melted honey or with *gummi Arabicum* dissolved in water. Powder it so fine as if one were to paint with a brush with it, or write with a quill. Then wash the honey off with distilled water; put the powdered ☉ into a glass bowl and pour warm, distilled water on it. Stir it well with a clean rod and let it settle down. Pour the water off above and add other distilled water; stir it again and let it settle down. Again pour it off, and repeat this so often till the water runs off as clearly as the one you pour on. Then it is enough. Now put it on warm ashes and let it dry, and you have a subtle powder. Now set your powder to reverberate in a reverberating—furnace in which glasses are made, twenty, twenty-five or twenty-six days, or till your gold swells as thick as a sponge. However, do not put it so hot that it melts, but keep it in a gentle heat without melting. Or if you wish, you may also dissolve your gold in *Aqua fort* and pour pure, distilled water on it; then let it boil for half an hour in a glass. Then put it aside for a day or two and your gold will drop to the bottom. Pour the water off cleanly, and again pour other, common, distilled water on it. Again boil it for a half hour as before, and put it again aside. Then it will settle at the bottom. Pour the water off. You may do that three or four times, till your gold is well washed off from the Aqua fort, which should be burnt of saltpetre and ✳. When your powder is thus washed, set it to reverberate, for this gold-powder which has been dissolved in Aqua fort does not melt so easily as the powder which has been rubbed on the stone. In addition, it probably takes at best ten days for reverberating; otherwise, both are equally good. You can therefore perform any of these two, whichever you wish.

When the powder has swollen like a sponge, it is sufficiently reverberated. Now take it out, have well distilled wine-vinegar and put the powder into a glass with a wide mouth, ground even above, upon which there should also be a likewise ground, round glass fitting the mouth. Now pour your vinegar upon the powder, so that two parts of the glass are full. Stir well and set it on a cupel with ashes. Close the glass above with the small round glass and give it also the warmth of the sun, stirring it every day, three or four times. Each time put the ground glass back on top, and your powder will gradually dissolve in vinegar, so that your vinegar will turn a very beautiful yellow. Decant the yellow vinegar into a clean glass and put it away well stoppered. Pour more vinegar upon the powder and stir it again. Set it in the furnace and do as before till your vinegar again turns yellow. Decant that to the first, and again pour fresh vinegar upon your powder; proceed as before. Repeat this till your vinegar is no longer colored; then pour it off, and take out what remains in the glass, dry it on warm ashes and set it again to reverberate as before, for eight or nine days. Then take it out and put it back into the glass. Pour distilled vinegar on it, stir well, and set it in the furnace, and do as before, till your vinegar turns yellow. Then pour it off to the first colored vinegar and again pour other vinegar upon it. Set it in the furnace and do as before till the vinegar is no longer colored. And if something worthwhile is left over, set it again to reverberate and proceed in everything as you have been instructed before, till all your powder is dissolved. Some feces will remain, because they had flown into it from the ashes in the furnace. The gold, too, has feces inside, so that some feces will always remain. You will yourself see if something is left worth reverberating or not. If there is something, proceed as indicated before; but if you do not think it worthwhile, let it be. But you can also keep those feces so that, should anything have remained in them, you can take it out afterwards.

Now take all your colored vinegar and set it to congeal on hot ashes, in an open glass vessel. A yellow powder will remain. Take it and dissolve it again in common water. If it does not dissolve, dissolve it again in vinegar and congeal it as before. After that it will dissolve in common water. It also happens that some feces stay behind. Put those together with the first feces. They are of no importance; they come from the innermost fecibus of the ☉. Now congeal again on warm ashes to a powder, and be careful not to give too much heat, for the powder would run together because it is "meltable". If you make it too hot, it will melt like wax. When it is then congealed, dissolve it again in common water as before; pour off the pure, and if some feces still remain at the bottom, they are of the innermost fecibus. Add those to the others and congeal again. Repeat this congealing and dissolving till you find no more feces at the bottom of the glass. Then it is enough. Then congeal again. Have a glass plate made specially for this purpose, or a marble plate, and spread it quite thin on it. Put it into a humid cellar and put a small glass underneath it. Everything will dissolve into clear water. Now congeal it again on hot ashes to a powder. Now it is ready to be added to the oil made of the Vegetable-stone.

CHAPTER XLIII

In this (chapter) I will instruct my son how he can blend the Vegetable-oil with the said gold-powder into an oil. Take therefore the vegetable-oil and weigh it. Take the same weight of gold-powder. Divide your *pulverem folis* into three parts, and put your vegetable-oil on fire, in its glass vessel. Give it natural warmth, as the sun shining in mid-summer. In it put the first third of your gold-powder; stir it with a rod of bostree wood so that it becomes well mixed. Let it thus stand in even heat for seven or eight days. Then add the other third of the gold-powder to the oil; stir it as before, and let it stand for another seven or eight days. Subsequently, add the last third of the gold-powder to the oil, and let it stand for another seven or eight days in the heat. Then everything is fixed, and a *medicine* for congealing ☿ into ☉. Heat a thin silver tin plate (or: griddle); when it glows, cool it in this oil and it will change into gold. Or take one hundred parts of *mercurius sublimatus,* which has been sublimated by gold ten or twelve times, until it left no more feces. Then *imbibe* one part of oil into one hundred parts of this mercurius; afterwards put it into the egg and hang it in the secret furnace for forty days and nights, and everything will turn into medicine. Thank God the Lord, my child. We shall deal in detail with these works in the Mineral-stone.

CHAPTER XLIV

Now then, my child, I have taught you the first part of the Vegetable-work. If now you wish to *operate* in the Vegetable, where the air first goes over, such as in honey and in the fruits of trees, or in wheat or other cereals, as also in everything where the air goes over first, it must be done in the way taught here. If you work differently, you will be cheated and not obtain the Vegetable-stone, and your work will be in vain. Follow this way, therefore, and you do not make mistakes.

My child should know that one can do many kinds of work of Vegetable-stones in the Vegetable, i.e., of herbs, trees, leaves, roots, seeds, wood, gums and other spices that fall under the *Vegetable*. All of them may be turned into a stone, which is altogether a Vegetable-stone. But they are made in two different ways. I have taught you the first way where the air goes over first, which work one must follow in all Vegetable—stones where the air goes over first, as has been taught here in an elaborate way.

HOLLANDUS'
MEDICINAL
RECIPES

FROM HIS

SECRETS

concerning vegetall and animal work

EXTRACTED FROM

THREE EXACT PIECES
of

LEONARD PHIORAVANT

1652

HOLLANDUS' MEDICINAL RECIPES

QUINTESSENCE OF HONEY

Now I will open to you a great secret in the Vegetall work of honie. To wit, a marvelous nature: for it is drawn out of the most noble and pure part of the floures. The nature of Bees is such that they draw out the best of everything as is enlarged upon in the Animal Work. Therein is taught how to extract the nature of all beasts, especially as in the 84th Chapter.

Wherefore my son, know this: That all that God hath created good in the upper part of the world, are perfect and incorruptible as the heaven. Whatsoever is in these lower parts, whether it be in beasts, fishes and all manner of sensible creatures, herbs or plants, it is indeed with a double nature. That is to say, both perfect and imperfect. The perfect nature is known as the Quintessence and the imperfect is known as the Faeces or dregs, or the venomous or combustible oil. Therefore, you shall separate the dregs and the combustible oil and then, that which remains is perfect and is called the Quintessence, which will endure continually, even as the heavens endure and it can neither be dissolved with fire or any other thing. For when God had created all things and looked upon them, they were all perfect good and there was nothing lacking to any; and therefore, for loves sake I say unto thee, that God hath put a secret nature of influence in every creature, and that to every nature of one sort or kind, he hath given one common influence, and to every one of several kinds, their several influences and virtues. This is whether it be on physics or other secret works which are partly discovered through natural workmanship. And yet, more things are unknown than are apparent to our senses. What? Do you not think that an herb is appointed for one disease which it will cure and also contains in it many more virtues than are known unto us? Yes truly, many more. I will add this as well: that if the Faeces and combustible oil be taken away from this thing, or herb, which in all things is the poison that should be taken away, that brings death to us, and the Elements should be purified and so burned together by Art, that they shall pass together by in a Limbeck and be joined together, as it were coupled in marriage, that it may root out all manner of disease from everything. This, be it herb or living thing, or be drawn from his venom, as in the 14th chapter of the Prologue of this Book is declared and also in the Prologue of the Animal Work. The manner of drawing the Quintessence out of all venomous beasts, birds, worms and flies is plainly declared, that it may help all the griefs of man, but that is specially drawn out of the blood of man, and there is likewise declared, that there is no need of things without man or beast to help such as are infected.

This is because every creature contains in himself the remedy of his disease. This remedy may be drawn out without hurting the man or beast, in order that the disease be miraculously cured as is most excellently taught in the Theorick and in the Practick. Therefore, I would write this, that thou might soon understand what marvelous force is in Honie, which is taken out of all floures and gathered into one Masse which is truly imbued with sundry virtues.

If God hath given unto other things the gift of healing, what then is there not in Honie, which is gathered from many floures and many herbs, and are all endued with a particular virtue? Truly if it be brought to his height and excellency, it will work marvelously. Now consider what lies hidden in this Quintessence and esteem it not lightly, but keep it secret as the most excellent thing of all Animal work. If this is obtained, you will need no other medicines to put away all accidents of the body.

CHAPTER II

Now I will set in hand with the practice. Take
twelve quarts of the best Virgin Honie and put it in
a great earthern vessel with a Limbeck well luted.
Set this in Balneo and lute a recipient to the neck
of it and distill that which will distill of it,
which is boiling in your Balneo. My son, know this,
that there is no common water in Honie, but only
Philosophical and Elemental. For the element of
Aire, does pass first together with the element of
Fire in which the Aire is contained. The air, when
it rises, resembles the savor of Aqua Vitae
distilled. Initially, it cannot be distinguished
from Aqua Vitae either by sight or by savor. Distill
it then, until no more arises, the leave the vessel
in Balneo five days with a Limbeck and receiver. Let
it boil night and day that the matter may be dried.
Cool it now, take it out and remove the receiver and
Limbeck. That which is in the receiver pour back
into the vessel over the dry matter. Set it back in
Balneo and cover the mouth of the vessel with a
clean, well luted dish, and let your Balneum be only
lukewarm.

My son, understand that it may thus be done, for it
is good that the fire be drawn with his proper air,
so as a man would stay so long, for it would be of

greater force. The ancient Philosophers wrought in this sort, but the danger is, when the vessels shall be opened, the water may fly away it being as subtle as wine. For every time the air is to be drawn away, and again to be poured on, making putrefaction in a warm Balneo, but first it must be well luted and a Limbeck being set on with a receiver, you must reiterate the work, until the fire rises like red blood. There is yet another method or rule of working found out in these our days, which is in this sort.

CHAPTER III

They are thus drawn out and the matter is dried, as has been said. Then take common water which has been twice distilled in Balneo and pour on as much as is sufficient and set in Balneo. Cover the mouth of the vessel but don't let the Balneum boil. Let it stand thus for three days and three nights, moving it day and night with a wooden spatula or spoon that is clean. After this, let it cool, remove it, pour it out and strain it. Then, take a clean vessel and decant the clear liquid and then pour on the Faeces fresh distilled water (rain water best) as was done before and set the vessel in Balneo as before. Let it be cleared and put aside with the first water and pour once more fresh distilled (rain) water and once

186

again set in Balneo. Do this as often as the water is tincted or coloured. When it no longer is tinged, you have separated the fire from the earth. Reserve the earth, or Faeces, until I tell you further what to do with it, for there is a combustible oil in it.

CHAPTER IV

Take the vessel containing the colored water and set it in Balneo with a Limbeck and receiver well luted. Distill all the water with a boiling Balneo and let the matter be well dried and cool. Then take away the Limbeck and let the vessel remain in Balneo and pour on again (from the receiver) the water over the matter and make a fire. Set a dish upon the mouth of the vessel and let it stand in Balneo three days. Stir this every day, three or four times with a clean wooden spatula. After this, let it cool, remove it and filter it. Then take a clean vessel and carefully decant the clear liquid into the vessel and right away pour on the Faeces fresh distilled (rain) water, stirring it with a wooden ladle and let it stand one day to clear (settle) and the Faeces that remain, put them in with the first Faeces which has been set aside. Then take a clean vessel and set it in a boiling Balneo until it is thoroughly dry and repeat this process until there remain no Faeces in the bottom of the vessel. In this way, you shall obtain the pure element of Fire:

and the element of Aire must also be so often distilled until there remains nothing in the bottom. This is the manner in which the pure elements are obtained. Separate then the water from the fire, and let it dry. This will give you a clear shining matter similar to Camphor. Keep the Fire well in a glass contained and the Aire with the Water in another container of glass, well sealed, until you have your earth prepared.

CHAPTER V

Take all the earth with the faeces and draw out the combustible oyle (oil) by a discensorie, that is, with two vessels joined together and luted (probably needs a vacuum) until the Combustible oil passes. This oil is useful for all cold diseases and other passions. If you do not want the combustible oil, just let it fly away. Then take your earth and calcine it in a reverbatory furnace, gently, until it be all white as snow. Then take a great earthen or stone vessel and put into it this white calcined earth on which pour a goodly amount of common distilled water. Stir it with a wooden ladle and let it stand three days in a boiling bath and keep it covered with a dish. Daily, stir it a dozen times. Let it cool, remove the vessel and let it stand to clear, for one day. Now, take another clean vessel

and softly pour out that which is clear (decant).
Upon the Faeces, again pour fresh distilled water
and once again set it in Balneo. Cool, remove, let
stand one day and decant into the first waters thus
obtained. A third time pour fresh distilled water
over the Faeces and repeat the entire process. The
Faeces can now be thrown away as they no longer
contain any value.

Take the vessel with these three waters and set it
in Balneo with a Limbeck and receiver. With a
boiling Balneo, draw out the water until the matter
be dry. Let it cool. Take away the Limbeck and pour
the water (from the receiver) on the earth again and
set it in boiling Balneo for one day. Let it
dissolve and clear. Decant that which is clear and
put in a little distilled water on the Faeces, and
let it stand for two or three hours in a warm bath.
Remove it from the bath and allow it to stand for
two or three hours and pour out the upper part upon
the first waters and the Faeces can be thrown away.
Once again set the vessel in Balneo, with the earth,
or salt, and distill away the water until all be dry
as before. Repeat this work until no Faeces remain
in the bottom. Drain away the water from the earth
and it will be like Crystal. Pure.

CHAPTER VI

Take a great glass that will bear the heat and put
into it your Fire and your Earth and pour your Aire
upon it and set it to distill in a furnace, in a pot
with sand or ashes, with a Limbeck well luted,
having a hole in the uppermost knottie part that a
funnel may be put in when there shall be a need of
Infusion. When as the humidity that it hath received
be half consumed, then fortify your fire a little,
gradually until you see the water start to boil.
Keep the fire in this state until the liquid boils
out so that only a pint remains. Remove the fire,
let the glass cool and take away the receiver and
open the hole in the Limbeck and put in a glass
funnel. (Note: Limbeck NOT removed from vessel).
Pour in all the water that distilled over into the
receiver. Plug the hole in the Limbeck and set the
recipient to the neck again and lute it well.
Distill again making the same observations and
practices as before. Do this ten times. The tenth
distillation being complete, let all pass together
as the earth is made volatile. So the Aire, the
Water, the Fire and the Earth will ascend together
by the Limbeck and be brought into one substance
which were in four. One together in nature and now
simple as the incorruptible heaven, yet are they not
fixed: but notwithstanding they are so coupled

together and so intertwined; that by no means can they be separated. They will continue now together as one body, forever; even as the Christalline and uncorruptible heaven, which notwithstanding, is compounded of the four Elements. What do you think of this, my Son? Cannot this Quintessence help every disease that now infects man through his most excellent temperature, whether it be in heat, cold, moist or dry. For all are in it that he may distribute unto every one that which is necessary; even as the heaven when need requireth, gives unto the earth all things as coldness, heat or moisture. And yet, it is neither hot, cold, moist or dry, but of one simple essence, and that imbued with such a nature that it giveth unto everything that which is necessary. In like manner, this is what this Quintessence does. Therefore my son, Rejoice! Give the Almighty God thanks which has opened these things unto the Philosophers.

CHAPTER VII

Now my son, if you would bring this Quintessence to even greater perfection, take a great circulatory or Pelican, that is a great glass that hath a great head similar to a Limbeck, and in the top of the head, a hole by which the matter may be poured in by means of a funnel. This hole is to be stopped. Out of the head comes two arms bending around into the belly. This permits that which goes up to descend again, through the arms back into the belly of the Pelican. This is the form of the vessel or Pelican, that distills one into the other.

1.   Take then your Quintessence and put it into a Pelican and set this into ashes. Better yet, put it into salt, prepared and dried. (like a sand bath) Regulate the fire so that it is like the heat in summer, the extreme heat. The Quintessence will rise like red oil and fall down again by the arms of the Pelican. By repeated ascensions, the Quintessence will become thick like wax or syrup. So much so that it will remain in the bottom eventually, and no longer ascend. At this point, fortify your fire so that the Quintessence will again ascend and descend. Maintain this heat until it again will not ascend but remains in the bottom. Make the fire even stronger that it will once again ascend and descend. Keep this same heat until it again rise no more.

2.   Observe this manner of augmenting the fire until the water be fixed and the glass turns red hot. This will take about twenty four hours all-together. If at the end of this time, the Quintessence no longer arises, it is indeed fixed and is brought into his highest virtue. Remove it

from the glass while still hot or it will become hard as wax when cooled and you will have to break the glass to remove it. For when hot, like wax it will become liquid. But when cool it congeals and pierces every hard thing, as oil does any leather. Its color is like a Ruby, and through shining like a Christal, it gives light in the dark, sufficient to read by. What do you think of this, my son? Are there not many strange bodies created by God? Truly he has imbued the Philosophers with no less gifts, for they that can look into the secrets of nature, shall see it to be an incredible operation. For this is gathered by Bees of the subtlest parts of all plants, trees, floures and fruits, and at that time when floures break out and trees bud. It is worthily called the Philosopher's Stone, for it is fixed and liquifiable as Wax and as the mineral Stone transmutes the impure metals, so does this one alter diseases.

Hereby it appears that this bears the bell among all the Vegetals; whereas it being yet in his grossness and impure, it is but of small value for any use in Physick by whatever means it may be boiled or skimmed off, but always retains his nature because it consists of all the fruits of the earth, plants and trees. Whereof one herb is hot, another cold, another dry and yet another moist, one astringent, one laxative, some corrosive and others venomous. So, diverse herbs have diverse qualities.

3.   It comes about that if it helps one disease, by and by it hinders another for everything works according to its properties when as is there made separation in the body. And of this separation is engendered blood and other humours. They are just like gunpowder in that so long as they sit still, there comes no harm therefrom. But if it be brought to the fire, it will at once demonstrate its secret nature and is kindled with a destructive fire. A fire which cannot be quenched with water, for the cold and dry, hot and moist, strive among themselves, a wind is stirred up that breaks all things near it. The same thing happens with Honie, that when it comes to the area of the Liver, it separates there and shows its nature to pass up and swell with wind. It is no surprise then, that the veins of the Liver can be broken by contention. When this occurs, Imposthumes are created in different places and causes such inflammations that the veins break easily. Although many highly recommend Honie, these are not Philosophers nor do they understand the nature of it. But when it is prepared as a Simple, fixed as Wine, then it is the most potent of Medicines among the Vegetals. There is nothing like it.

Give God thanks and be generous to the poor. The dosage of this is one grain and it must be taken morning and night on an empty stomach until the disease is gone. Now Praise God.

CHAPTER VIII
ROSE SOLIS

Diseases of the Eyes, Rheums, Inflammation, Diseases of the Heart, Wolf, Inflammation of the Liver or Stomach; drives away dreams and fantasies, good for bites of venomous beasts, against poison that has been drunk, for pestilence, muscles, tendons, wounds and other ailments, and Canker.

Now my son, I will teach thee the greatest mystery or secret amongst all vegetable things, whose force and strength has been kept secret amongst all the ancient workmen, and they have bound themselves, one to another by Oath, that they should not utter, in their books or their writings, the strength of this herb which is called ROSE SOLIS, and in the German tongue: SINDAWE. Whosoever hath not the whole vegetable work, he cannot attain to the strength of this herb. For in that work is comprehended all the force of medicinal things. And this work of vegetables is not come to the hands but only of the ancient sworn Artificers, which were skillful in the liberal Arts.

But now my son, I will open it unto thee with adjuration, that thou shall keep this hidden knowledge secret. First, my son, you must understand, that this herb is the herb of the Sun, upon which the Sun spreads his beams and influences as he does upon gold in the veins of the mines; and he pours out his influences more upon this herb than upon any other herb which is created of God As it is evidently known to the ancient Philosophers, this herb far surmounts all other herbs which spring out of the earth just as the sun surmounts all other planets in the heaven, and hath greater force and power of influence than any other thing created of God in the firmament. So this herb excels all others in virtue and therefore this herb is arrayed with another color, other leaves and stranger shape than all other herbs. And his nature is such, that the hotter and drier the country is in the time of the year and the heat of the Sun, and the more that the Sun doth heat and burn him, this herb is the more moist and filled with dew So much so that upon one branch will hang a thousand drops of dew. As a test, strike this herb with a slender twig so that the drops which fall from it will fall into a large glass vessel and you will see it filled up with a marvelous dew. Now if the Sun is extremely hot, those branches will, in less than half and hour, be filled with more dew than ever before!! And if you strike the branch twenty times with the rod, each time it will be once again laden with dew. It is almost enough that we see no other miracle than this, where the dewey humor arise in so short a time even though the Sun scorches up all the other herbs. The hotter it is, the more moist will this herb be, as if sprinkled with water. Hereupon may we gather his marvelous qualities and judge that there is some secret operation hid in it. If you will keep this water, thusly gathered, in a glass vessel, you can with it cure all the diseases of the eyes whether they come of Rheums or of inflammations. It helps all the pains and diseases of the heart, it cools the liver and the stomach that is inflamed and mitigates all the pains of the head that comes from heat. It drives away all dreams and phantasies and is good to kill the Canker and the Wolf. It is useful against the biting of venomous beasts, against poison drunk, if it be taken by the mouth. It is also helpful in cases of pestilence and it is good for many other diseases and ailments as well.

CHAPTER IX

This herb has the color of the Sun, for his color is dark red, divided with yellow lines and his shape is

like a star. His proportion is like a heavenly
Planet and consists of seven branches.

Afterwards, take a large glass curcurbite and put
all three of your elements in it and set it into
ashes with a Limbeck and Receiver fitted and luted.
Make the fire in the furnace gentle at first then
increase it gradually until the fire and aire be
passed and the Limbeck turn red within. Then make
the heat moderate until all the element of fire be
passed and the head becomes blood-red and the water
and aire shall swim upon it like oil. In this way,
the three elements are brought to their highest
essence and are perfectly rectified.

Take away the receiver and stop it well, until your
earth be prepared. Realize that in the dust and
Faeces there remains yet a combustible oil which can
be extracted by a discensory, if so desired. It is
good against the cold-Gout, for members that are
numb and sinews that are too much mollified. If thou
be weary of this labor, put this powder or Faeces
into a reverberatory that they be mingled with a
gentle fire until it becomes white as snow. When
this is done, put it into a large stone curcurbite
and pour on it a large quantity of double-distilled
water. It matters not how much you pour on. Stir it
well with a wooden ladle or spoon five or six times,

always recovering it well. After four days, allow it to cool and let it stand four days and settle. Decant the clear liquid carefully from the Faeces and into another clean vessel. As before, pour on (fresh) double-distilled water and stir it with a wooden spoon. Set it in Balneo for two days then allow it to cool and settle. Then decant the clear liquid off and combine it with that first obtained thusly. Repeat this operation for the third time, then throw away the Faeces as they are no longer good for anything.

In the outerpart it is broad but near the ground it is narrow. It appears to be as if it were a heavy, tender substance, outwardly hot and moist, inwardly cold and dry. The left side of it is cold and moist and the right side hot and dry, and it is most temperate as Gold. Wherefore his Elements cannot be separated one from another as in other herbs, but it may be purged from his Faeces, for his fixing letteth that the Elements cannot be separated, for the fire will ascend with the air by the Balneum as we will hereafter teach. The earth may be separated from his Faeces, and the Faeces likewise from the fire and air, although it does not have many dregs. Some Latin writers call it LINGUA AVIS or Bird's-tongue; some call it SOLARIA, of the Sun, of LUNARIA it is called the Moon; the Fleming calls it SINDOW.

The old Philosophers have kept secret the qualities as yet for the marvelous effects that it works. And it is a marvel, says Arnold Villanueva, that a man should die that every day eats some of it in his gross substance. What will it then work when it is brought into his fineness and cleansed from his Faeces? It has this great virtue in it that if it is put into a glass where there is poison mixed with wine, or in any other cup where there is poison, the glass will instantly shatter!

If the container is made of stone, or alabaster or the like, the wine will proceed to boil vehemently as if there were a fire underneath the container. The wine will then run out of the container until nothing is left. Also if anyone carries this herb with them and comes across an enemy, not only will the enemy not have power over the individual, but must, in fact, serve the carrier of the herb. If anyone is bewitched in body or in his art, that is in his Cookery, Brewing or Baking, or by any other means, this carried on the person will set them free from the witchcraft. If it is tied upon the belly of a woman who is pregnant, the woman will immediately be delivered even though the baby had died and was rotten within. The herb when carried on the person, or when a little is eaten daily, as it is being used, it will prevent the occurrence of the Falling Sickness.

Further, if a person suffers from apoplexy such that his mouth be drawn aside and he is incoherent, his senses will be restored if the juice of this herb, which has been strained is administered orally. If the herb is hung about the neck of one possessed, the person will be still as a lamb and the power of the possessing spirit will be taken away. Prove it for yourself and you will find it is indeed true!

Bleeding from the nose is stopped if the herb is held in the mouth. Those who are wearied from travel, if they will take some of the juice in wine, they will before long be refreshed as if they had never taken the trip or performed and labors. It comforts the sinews and muscles, the tendons and all of nature. Also, it will heal wounds if taken for ten days in wine or ale and if the wound is washed with the same mixture and bound with a cloth dampened in the same. Toothaches can be alleviated if the herb is placed next to the teeth.

All these things have been done many times and proven effective. Consider: if it will do such marvelous things while still encumbered and weakened with its own gross matter, what wonders will it not work when brought to perfection? My son, know this for the truth, that there is no herb that grows on the earth that can compare to this in strength and effectiveness. Therefore, make sure that you don't neglect it but be mindful ever to keep the secret from those that are not of the nature of children and from the ignorant. For if this herb were to become scarce and its properties were to be made known to all, it would become more highly prized than gold or precious stones, for the effects of the Quintessence are marvelous as you will see.

CHAPTER X

Now it remains that we teach how this herb may be brought into his highest degree and to his Quintessence. First, it is to be gathered, the same having his course in his own house. That is, in the Lion and the Moon behold him with a sinister quadrate aspect Pick the herb whole along with roots, leaves and flowers and see that no dirt or earth cling thereto and that no other herb be mixed with it. It is also very important that the herb does not get wet or moist in any way. Therefore, pick it not when it rains but, rather, when the Sun is shining brightest.

Gather a large amount of the herb and pulverize it
well in a mortar made of marble and put it in a
Curcurbite made of Stone with a head and receiver
luted and set it in Balneo. Let all the water
distill away until the herb is dry like powder.
There will rise together with the water, the color
of fine gold. This happens only with this herb and
with no other. Now, when there is no more liquid
leave the curcurbite sitting in the Balneo for three
to four days and boil it night and day, so that all
the moisture is completely separated and drawn off.
Then let the vessel cool and take away the receiver
and stop it carefully and then take off the head
Take out the matter (faeces) and grind it well in a
marble mortar that it is fine enough to pass through
a coarse sieve. Put this powder in an earthen
curcurbite and pour on your water and aire and stir
it with a wooden ladle. Cover the mouth of the
curcurbite tightly and set it in a warm bath for
nine days so it may putrefy. Stir it daily with a
clean wooden ladle, four or five times, then recover
the curcurbite, weighing it down with a weight such
as a lump of lead. After nine days, take the vessel
from the Balneo and strain that which is in it into
a glazed earthen vessel. Strain it well so that the
matter dries. Then, take this dry powder and put it
into his curcurbite and cover it and keep it in a
warm place until you are given further instructions.

1. The moist liquor which will be drawn from it will be red in colour for the element of fire is there present with the air and the water. Put that liquid into a curcurbite of stone and put on a head and lute it well. Set it in Balneo with a receiver well luted and distill away all the water, with the aire severally from the fire, in a boiling bath until no more comes over and the fire will pass away in the bottom. Then take the vessel out of the bath and stop it well. Further instructions what to do with this will follow.

2. Then take once again the stone curcurbite wherein is your powder and pour on the fire and the aire and stir it well with a clean wooden ladle and set it in a warm bath for nine more days. As before, cover it well and stir it daily four or five times with the ladle. After nine days, strain out that which is in the vessel, and pour the liquid into a glazed vessel. Put the residue of the powder into another vessel as was done before, and just keep it there until you have the instructions on how to remove the combustible oil.

3. Take the vessel now wherein is your fire and mix your liquor with it which you keep in the glazed vessel where your fire and aire is, and set a head

on the vessel where the matter is and lute it. Place it in Balneo and set a receiver to it, to the bill of the head, then distill out the water and air with a boiling bath until no more comes over and you will then have in the receiver, water and aire. Take them away and take the vessel out of the Balneo and you will find remaining in the bottom a thick Turpentine-like substance. This substance is the element of fire mingled with much Faeces. Now the fire is to be separated from the Faeces in the following way: pour on your water and aire upon that whence you drew it, and stir it with a spoon and cover it with a tile and allow it to settle for four days and the Faeces will fall into the bottom. Carefully decant the clear liquid into a clean vessel (curcurbite) making sure no Faeces are poured over and stop the first vessel and set it by. That vessel which contains the water, fire and aire, set it in Balneo with a head and receiver fitted and well luted and distill the water and aire in the same degree. When no more comes over, take away the receiver: Let the vessel cool and you shall find in the bottom, your fire; which keep in his vessel well stopped. Then pour on your water and aire on the Faeces again, from whence you did draw them, and stir with a spoon and let it settle for four days. Carefully decant from the Faeces into the vessel that contains your fire, insuring no Faeces passes over. Cover your vessel wherein are the Faeces and set it by. But your vessel, wherein is your water, aire and fire, set it again in Balneo as I showed you and distill away the water and aire for they do always pass over together, and the fire will remain in the bottom. Pour on the water and aire again, upon the Faeces, stir it with a spoon and let it settle once more for four days. This is now the third repetition whereby you have separated the water and the aire from the faeces. That which is clear, decant into the vessel that contains your fire. Then put your Faeces into the first vessel which I bade you keep, in which your combustible oil is.

CHAPTER XI

Now you have drawn your fire out of your Faeces; therefore distill your water with the aire and when it will distill no more, take away your vessel and you shall find in the bottom, the element of fire. It is not yet pure but foul and full of dregs. Therefore, pour on the water and aire again and stir it well with a spoon or ladle. Cover it and let it stand and settle four days. Then decant out the clear liquid into another vessel (clean) and set aside the vessel wherein are the Faeces. Now, the vessel with the fire, water, and oil, set into

Balneo to distill so long as something comes over, then remove it and pour the liquor distilled into the vessel that has the Faeces and do as you were taught to work with the Faeces until you have your element of fire without any gross substances.

When you have distilled away the water and aire from the fire, and allowed it to settle four days, and that all is clear without any Faeces, then you shall have your pure fire. Therefore put all your faeces together with the first Faeces where the combustible oil is.

All the waters that you have drawn, distill by Balneo. To hasten the process, distill in ashes that all the water may rise and that there remain in the bottom a dusty matter. Pour on fresh distilled water again, stir it, keep it in Balneo for 24 hours, then let it cool and settle. Pour out (decant) the liquid that is clear, gently, from the Faeces. Pour some common water on them and stir it. Set it in Balneo for twelve hours, take it out, let it settle and pour the clear liquid into the first water and throw the Faeces away as they are worthless. Repeat this work often enough so that no more Faeces will settle out. Then you shall have your earth rectified from all his Faeces which you shall congeal or dry until it be like a powder or dust. Then join that with your other elements in the glass and it will at once resolve into his element for the element of water is there present. Set them all together in a furnace, upon ashes, put on a head with a receiver well luted. The Head must have a hole in the top so that liquid may be poured in but this hole must also be well stopped.

Make a gentle fire in the furnace, at first, but hotter afterwards, until that which is fermented does pass. However, do not draw out all that is in but about half a sextary of the liquor with the Water, that the matter may remain moist. If you were to drive it all out, it would congeal into a hard mass and break the glass in the furnace! Then, open the hole in the Head, and with a funnel pour in that which is in the receiver. However, warm it first to avoid breaking the glass by pouring in cold liquid. Of course, you can always first cool the glass before pouring. Repeat this imbibition ten or twelve times. After this, distill out anything that will come over as long as it passes through the neck of the Limbeck. After this tenth distillation, the earth will no more be congealed, but will rest in the bottom like a red golden oil. Imbibe it again, pouring on the liquor and distilling it until all the elements pass by the Limbeck and nothing remains in the bottom of the vessel.

CHAPTER XII

Give thanks now to God for His marvelous gifts which he has distributed amongst His Philosophers and hath given them so great a knowledge of things as they are uttered in this work, which is all together heavenly and more divine than human. For it is a great marvel in this life, that man's understanding can bring these inferior things to so great perfection that they have attained to the highest degree of virtue. Truly it is the work of the Holy Ghost, which hath put it into the minds of men. For I do affirm that whosoever hath this herb so prepared, that he may help all the infirmities of man's bodies whether they be curable or incurable, except natural death, which is ordained before unto every man of God. Yea, this I dare be bold to say, that if a man uses the weight of one scruple of this Quintessence, or the Quintessence of Sugar and Potable Gold, wherein pearls are dissolved, or the Quintessence of Selandine, that man by Gods help will not die before the day of the Great Judgment. For the humors in man's body can by no means predominate one over another, as is taught in the thirty six chapters of Vegetal. In here is treated the Quintessence of all Medicinal things and in here it is showed that by Gods help the life of man may be prolonged even until that day, void and free from all diseases and sicknesses. Further, man may be preserved in the state that he was in at thirty years of age and in the same strength and force of wit. On this all the Philosophers agree that a man may continue in the same state as long as in an earthly paradise. This is so plainly shown in that chapter, that willy-nilly, you will be constrained in your mind to believe it and to admit that it is true.

Therefore, it is not necessary to reason much of the force and quality of this Quintessence, but whatsoever a disease man be infected with, give him as much of this Quintessence as a nutshell will hold, in wine. In a short time he will be cured as if by a miracle according as the disease is gentle or violent. But if you give this Quintessence to drink mixed with Quintessence of Sugar, with Potable Gold, wherein pearls are dissolved and with the Quintessence of Selandine, within one day you will cure all the diseases and sicknesses whatever they might be. It can be seen by this that the work is divine rather than human. Therefore, give God the praise and take heed that you do not reveal this secret. For by this means, tyrants would prolong their lives so as to accomplish their wicked deeds and purposes (whereof both you and I would be the occasion). So then, keep it secret, for it is one of the greatest secrets amongst all the Vegetals. There is no treasure that can compare to this work. Perform this Work then, and distribute it liberally amongst the poor and God will give thee eternal felicity.

CHAPTER XIII
QUINTESSENCE OF SUGAR

Here will I show thee a great secret, how to draw the Quintessence from sugar. This truly excels all the vegetable works by means of his temperature like the incorruptible Heaven which is never hot, cold nor dry; but most temperate but nevertheless compounded from the four Elements. But these do not strive with one another for they are so conjoined that they can never be separated. They remain ever simple and fixed in their unity. But this Heaven does distribute and give unto the earth whatsoever is necessary for it, although that itself be neither hot nor cold, moist nor dry. The Quintessence of sugar has the same Effects and contains the four elements such as does gold. As gold is pure, so sugar is impure; gold is outwardly hot and moist, inwardly cold and dry and white. Sugar is just the opposite, for it is outwardly cold and dry and inwardly hot and moist and red. Further, it is fixed both inward and outward. Nor is there anything wanting but that the inward quality may be brought, that his redness may appear outwardly and that his Faeces be separated. Then it is prepared and does not need to be fixed for it is fixed already and retains within itself all outward and inward spirits and all that is volatile.

Now what his kind is, I will tell you, even from where the original came, that is, even out of the red. However, for more information, read the thirty-three chapters of the generation of those things that grow in the Seas and other waters whose nature we write about in detail. Here it will be sufficient to show the order how to prepare it and in what order it should be used for Medicine, also to what other things it might be applied. His nature is to retain or hold all flying (volatile) spirits and to fix them into a stone as shall be shown hereafter.

CHAPTER XIV

First you must understand that you cannot separate the Faeces except that you bring the inward parts outward. That is to say, that his inward dark, golden color must appear. When this is distilled, then the redness will be seen and this fire, passing the yellowness of his aire (his incombustible oil), then you can first separate the Faeces from the

Quintessence. Take then, hard and white Sugar, for it is not necessary to travel much in dissolving and coagulating it even though there be much impurity therein, that hinders not, but that the inward part may be brought forth, for it must be purged when as the redness shows outward.

Take therefore, ten to twenty pounds of Sugar, more or less as may be convenient, and pound it fine. Put this into a curcurbite of hard stone and top this with eight fingers of Aqua Vitae. Then distill it in Balneo with a strong fire until nothing further comes over. Let it cool and pour on the said Aqua Vitae again. Repeat the process six or seven times. Upon completion, open up the head and take out the sugar and place it in a strong glass and set it in fine sifted ashes and pour thereon the Aqua Vitae and distill it until half the liquid (Aqua Vitae) comes over. Then pour this Aqua Vitae back on after warming it so the glass will not break. Note that the head should have a hole in it to permit the pouring on of the Aqua Vitae by means of a funnel.

Repeat this so often in a strong fire that the wine and the sugar may boil in and because the half part of the wine will come away quite rapidly, you must at once put on the other part. For if you should distill all the wine out, the Sugar will burn due to the heat as it must be continually boiling in the glass. (i.e., always keep half the liquid in the distilling vessel to prevent burning) Also, it would smell of the burning because of the incombustible Sulphur in it. When you observe that half the wine is distilled forth, warm this half and put it in the vessel again with a funnel. Repeat this process often enough so that the Sugar remains red as blood as can be seen through the glass. This procedure will require eight to ten days of effort and is dependent on how you attend the fire. (The text is not clear if the work should be done in an uninterrupted way, i.e., non-stop, no sleep, etc. and this will depend on how fast it distills over).

When the matter does turn red, let it cool and remove the vessel with ashes and set the vessel in Balneo and with a strong fire distill off the Aqua Vitae until the sugar remains dry and when it will distill no more. Allow it to stand very hot, in Balneo, for about four to five days. This will permit the sugar to perfectly congeal. Then let this matter cool and remove it (this 'stone') which will be pitch black. Then take this stone and put it into a great quantity of double distilled common water and set it in Balneo for five or six days, with a great heat, lightly covered. Stir it daily five or six times with a clean wooden spatula. Let it cool.

Remove it and allow it to settle three or four days. Decant the clear liquid into another vessel and close it well. Then pour onto the Faeces the sublimed water as before, and set it into Balneo to digest for several days, stirring it as before with a clean ladle. Then cool it, settle it and decant the clear part to the other part already decanted. Then pour more water on the Faeces and digest it in Balneo as before, etc. Do this so long as the water contains color (tincture). Once it is no longer tinged, then stop and cast the Faeces away as they no longer have any virtue in them. Now the element of earth is with the element of fire and water, neither can they be separated anymore but are fixed together.

CHAPTER XV

Make a trial thereof by burning some of this substance. Nothing will remain but, perhaps, some light ashes. It will burn like oil or fat. Now, take the glass wherein is all the red solution and distill it in Balneo or let it simply evaporate, if you have any more of the distilled water, until it be dry. Then, let it cool and take off the head and pour thereon more sublimed water and set it in Balneo again. Stir it with a wooden ladle as before and let it settle as before and decant the clearest part and do this until there are no more Faeces.

Then put it into a glass that can bear a great heat and boil it away or evaporate it until a certain scum appears on it. Then, take it forth and set it in a cold and dry or hot and dry place and it will grow into a great mass or lump, red in color and transparent like a Ruby or other Philosophers Stone which if you will reduce it to powder and set in a dunghill, in a large wide glass, allow it to evaporate and it will come into a yellow powder like gold. This then is the fixed Quintessence of Sugar which retains all volatile spirits. Nor will this be sweet but have a heavenly taste which when put in the mouth will melt without any feeling. If it be winter or cold, you will notice a heat naturally pervading the body and a feeling of lightness that makes you seem to be able to get up and fly! If you become too warm, swallow a little and you will soon cool down as if in a cool bath. Thus it works, in heat, moisture, cold and dryness, by an incredible Miracle.

When you wish to use it, drink it with rectified Aqua Vitae, or Rose Water, Endive or Scabios, or by itself, and you will witness Marvels. If anyone be diseased outwardly with scabs or ulcers, let them drink of this and wash the sores with wine, wherein the Quintessence is dissolved, and, like a miracle, this person will soon be cured. If anyone is wounded or stabbed with a weapon, so that it is not lethal, let him drink a drachma (3 i) of this essence with warmed wine and wash the wound with wine wherein the Quintessence has been dissolved. This one shall be cured in an amazing fashion. It helps in the case of falling sickness and in pestilence and all such diseases as may happen to man. If you possess the Aurum Potable, mix 2 pounds of this with a pound (lb i) of the Quintessence in a glass vessel and set it on a trivet or a dry Balneo thirty days in an Athanor and they will be mixed together. Then they will most assuredly work miracles in mans body. Further, when you have extracted the Quintessence of any herb, coagulate and mix it with some Aurum Potable for further miracles. Now, if you will have it pass the helm, you must put on as much vinegar of Aqua Vitae and distil it. Again pour on fresh vinegar or Aqua Vitae and draw it away again until the Quintessence ascends in a red-golden color, as pointed out previously in several places on how to distill those matters that are fixed by vinegar or Aqua Vitae, for when it is distilled by Limbeck, his virtues are magnified a thousand fold and will work unusual cures. Keep this as a secret for it is a great mystery in nature.

EPILOGUE AND COMMENTS ON THE PREVIOUS TEACHING

The natural, earthly man is so much afflicted by nature with strong emotions that he feels them in all circumstances and is almost never found in an impartial frame of mind. That, however, is quite incongruous with true wisdom and also altogether contrary to the Christian teachings, as is expressly specified by the Apostle James when citing the characteristics of wisdom with the words: Wisdom from above is first of all chaste; after that, peacefully modest; let me tell you that it is full of mercy and good fruit, impartial and without hypocrisy. Such partial hearts come to the fore especially when something unusual happens to them, when they make too much or too little of things, so that even the most highly esteemed, when it comes to this, are easily overcome by a little female passion.

Of that we also have the example of an untold number of persons in regard to this author. One party, on hearing about such an excellent man, falls for him, almost making an idol of him. The other, on the contrary, cannot get it into its head but stands up against it, calling it vain lies, cheating nonsense, and bragging. Both, however, are going to unfair extremes, and the middle course would serve them better (that is) to examine everything thoroughly, without emotion, and by the good found, recognize the giver.

If great talents are found in someone, he has certainly not got them of himself; but if someone is full of stupidity or deficiencies (shortcomings), he can by the same reasoning not be better of himself. Each (of us) should always remember this. In adversity it will stand him in good stead. In addition, it should be highly necessary and unforgettable for any conscientious person, namely, (that) if he sees anything specially good in any person, he must never praise him in his presence, so that he does not become annoyed and thereby tempted to think a great deal of himself. On the other hand, if he becomes aware of someone's fault, he must not diminish him in his absence or bring contempt upon him, speaking in his heart: I thank God that I am not like other people, etc.

Now then, in regard to what I wanted to remember of our above mentioned author Hollandus, it is concerning the excellent and exceedingly great *arcana,* on which he gives information in all his Vegetable preparations, that I intend this time to put my understanding down here, what one is to think of such high matters. (I am doing this) for the sake of some of my co-disciples who are beginners, and who have so much innate intelligence that they recognize that their love for this splendid study cannot be of use to them unless they have previously managed, through untiring industry, that they can, as far as these secrets are concerned, look into the hearts of all old philosophers. Because of this they have afterwards the advantage that they cannot harm either themselves or others in their *practice*.

Instead, others who despise such means of diligent reading, wishing to obtain great experience without it, are often punished by being obliged, after spending their own funds, to look from time to time for other sponsors, in order to test the processes they devised in the laboratory - until they finally completely despair of the art. Accordingly, I find in my understanding, which is likewise still at the first stage and eager to learn more, that everything that originates in Divine Creation is pure power and might of God the Most High, the visible as well as the invisible; and nothing created can be found or devised which is not either a substantial, tangible, hard and dry, or soft and liquid, or else an invisible, intangible, spiritual power. All such powers, no matter how innumerably—manifold they be, have their root and origin in the *Mysterium Magnum,* which is the might for all such powers, and proceed from there *de potentia in actum.* Thus it turns into an innumerable—manifoldness which yet arises only from and lives in one single root. Just as may be seen that the many kinds of plants, whether they be hot, cold, sweet, sour, poisonous, salubrious, or whatever their nature, have all of them their life and growth from the one sun. And when fall and winter take the sun from them, they must all die, although they have contradictory properties and are yet of one life.

Such it is also with all powers created *in rerum Natura;* in their mother and origin they are but one and are therefore infused into such opposing properties that one should reveal the other all the more. For how could one judge a sweet taste if the sour or sharply salted were not known to him? How could one truly recognize what is delightful without the harsh and bitter, etc.? Since then such opposites come from one ground, and it is one and not two or more and therefore no discord can be in the one, for it is only one and not more and has no opposite, one may conclude that such opposition is or arises only in manifestation. Likewise, when it returns again to its beginning or extreme end (one may conclude that) it is no longer so.

In this world, however, such opposites exist in all things, because God, for the revelation of his infinite wisdom, has ordained that Nature shall not cease one moment bringing forth varied colors, powers, virtues and wonders. This manifoldness requires that it must make one thing hard, another soft, a third cold, a fourth hot, a fifth dry, a sixth humid, the seventh dark, the eighth bright, and the like, through its strong activity. Those properties are then also easily changed one into another, as also overcome one by another; just as when air turns into water, and water again turns into air. In the same way darkness is illuminated, brightness is darkened, heat is cooled, coldness is heated, dryness is moistened, moistness is dried, motion is stopped, and motionlessness is moved, and what else there is.

From those accidental and transformable properties one may recognize with all philosophers that the same applies to the human body. Through constant putting in motion and circulating of its vital power, frequent changes arise in its *properties*. Sweetness sours, purity becomes obstructed by mucus, the temperate becomes hot or cold, and (there arise) countless more happenings that cause sickness and death. As there is one thing in all growing things, however, which makes that in it they are one in all their contradictory disunion; that all of them take the spark of their life and growth from sunshine of the right temperature, and are at peace in (better: are identical in) this, no matter how unlike they appear in color, power and virtue. Thus one may also recognize that the same thing is one with the sun and is the life of all things; but it must be ignited by the sun, because in all other things it is locked in too hard. But in the point of the sun life is manifest, and from there it must also be excited in all other things.

It must not be thought that the sun and planets are only in the sky. They are everywhere through the All as seven spirits or qualities, which have been noticed to follow one another every 24 hours in planetary *operation* at the edge of the created world. Thus then is the same one life the *point* in all things and has been called from old *Quinta Essentia, Mercurius Vitae, Tinctur Physica, Avis Hermetis, Lapis Animalis Vegetabilis & Mineralis,* and what other names it may have. In itself, in its root, it is no other than living Sulphur and must with its like always be ignited by the sun. When we human beings lack the warming sunshine in winter, we must warm ourselves with an earthly Sulphur—fire, which burns only in wood, peat, coal, etc. And whoever is deprived of such warmth in great cold, his members and body first begin to twitch, finally turning quite numb. Nor could anyone live where such Sulphur-warmth were greater (than the right temperature for human beings). But why (must) this evanescent *tincture* or life—ignition be so easily obstructed that it must stop tincturing its *Corpus* with life and motion, which is death for the creatures of this world? That has been caused by the envy of the abominable Satan, for God did not make death. Neither does he find pleasure in the destruction of the living, as will be proved in lib. Sup. cap. I.

Whoever knows how to draw the Quinta Essentia out of where it is and rid it of all *fecibus,* gets with it a real bodily sunshine which, on account of the *concentration,* will strongly prove the whole might of the sun in a small particle and is not, like the sun of heaven, again removed from the possessor's horizon. Therefore, such real sunshine is to be sought most, next to God.

The kindhearted philosophers did not neglect to most diligently leave to their successor's inducement and teachings for reaching these gifts of God without fail. Of that it is not necessary to bear witness in regard of *Hollandus.* His own testimony is more important than that an intelligent person should not accept it. While we have understood from his words that he can cure with the Quinta Essentia all diseases and infirmities that came to his attention, including ridding possessed persons of evil spirits, I have already admitted before that I cannot say anything else about it but that it is divine truth that in the true philosophical Quinta Essentia, prepared out of the pure fire of the sun and cooked in the dew—water of the moon, such heavenly, supernatural might was not only found by Hollandus but also by other illumined philosophers.

Just as the author of the great *Rosarium Philosophorum* states publicly that all illnesses that befall the human body, from the crown of his head to the sole of his foot, can be taken away completely by the philosophical tincture; even if an old man uses that tincture, it can make his senile hair fall out and other hair of his previous youth grow instead, and restore youthful vigor and strength. *Basilius Valentinus* exclaims with great affirmation and stating that he would answer for it on Judgment Day, that in the *Aster Solis* the power and effectiveness of all other *subjects* are concentrated and may be obtained gathered together in it, the whole *Medicinal operation,* and much more, as all other plants, stones and minerals can prove.

He also gives more than one example of how he himself cured extremely painful bladder stones which, however, many consider incurable.

This is confirmed by the pious Count *Bernhardus* (Trevisanus) by writing in his preface to his *chemically* true booklet: "Let no one grudge the labor, or even regret it, while it is known for sure that by it he can escape intolerable poverty and all infirmities of mind and body: Since I myself have *experimented* and helped people troubled by leprosy, epilepsy, dropsy, consumption, strokes, and gout; also those who were possessed by devils, who were raving and insane, and many others."

In the same manner the philosopher *Trismosin,* preceptor of *Paracelsus,* writes in his "Treasury of the Red Lion" as follows: "Man cannot speak of this secret, much less think of it. This is the reason why it is the greatest treasure in this world that may be given to man. And if GOD the Almighty LORD of heaven and earth would help, man might live and sustain his life for four hundred years with this *arcanum* when it turns into the *Medicina.* For the great fire of this secret renews man from scratch, so that the *humor radicalis* is totally renewed in the human body. And I, Trismosin, say by my highest truth, that I have given of this medicine, as I had prepared it from the red lion, to 60 and 70 year-old women who afterwards bore children again. I gave of this medicine to an old man of 89 years. He became transformed (younger). His skin and hair all changed, and he lived for thirty more years afterwards."

Enough such testimonials are found with many other credible authors, and I have only quoted these few so that the beginning seeker should feel assured, aside from me, that God has provided for superabundant help for all our infirmities, not only of the soul but also of this wearisome body, provided we seek understanding from him in long-lasting, earnest persistence, so as to partake of it ourselves. But whosoever craves or desires understanding and true wisdom must know that such is no other than the breath of Almighty God or the breathing of the Divine Power and the effulgence and radiance of the Eternal Light, which the wise author of the splendid Book of Wisdom loved above all treasures as a most noble, chaste virgin, forever abiding with God. And he asked the Most High to give her to him as his bride, who was then also united to his soul in an eternal marriage bond. He testifies, however, that she does not enter evil souls at all, nor dwell in lives subject to sin, that is, of those who wallow in all kinds of sinful mud, such as gluttony, drunkenness, whoring, lying, cheating, arrogance, etc. That is why each must get rid of such monstrosities as well as of all tempting and bad company. He must be a complete transformation, become accustomed to a penitent and pious life for as long as he lives, praying to GOD day and night for the spirit and mentality of Christ. Then he will acquire the precious pearls which all other sages have also possessed. And when knowledge is granted thus by the Father of Lights, the longer it lasts, the more he will understand from where the opposites in nature come, and how one put opposite the other causes a struggle, and how one drives the other away; also that there is as much *potentia* in one as in the other, but that *in actu* one is at times superior to the other, just as a greater fire dries up a smaller moisture. On the other hand, if moisture or water is present in a larger quantity, it extinguishes the blaze of the fire. Aside from that, however, there is in nature as much *potentia* or ability in one as in the other. One whole element cannot be or become more than its opposite. The whole element water cannot dissolve the element earth or *predominate* it. Likewise, earth cannot congeal water or make it thicker, and likewise, *compariis potentialiter* with the others.

In actuality, however, it is as follows: *Gen unius est corruptio alterius,* since now *vie coagulativa* now *vie solutiva* predominates; now *subtiliativa,* now *incrassativa,* etc., as the *Philosophia* proves in several ways, while also teaching how to overcome one of these contingenci with another.

When a *Tartarus,* or stone, *coagulates* in a person, it must be *reduced* by means of *vis solutiva* and again dissolved. Similarly, how to bring the opposite qualities into *temperature,* so that one is in balance by the other and none overpowers the other or can itself be overcome. That is the content of the teachings of all philosophers concerning the *Quinta Essentia* which, when it is brought to its highest degree of perfection, can no more be overcome at all. Instead, it can do as much *in patiendo* as fire or some other things *in agendo.* For this reason *Tauler* says of a perfect human soul that it is all powerful in suffering, just as God is almighty in action.

Since the Quinta Essentia is an indestructible *substantial* life, in which all opposites are uni and brought into one simple mode of existence, also makes everything *temperate* and in balance it reaches man's body. If then *vie coagulativa* vanted to predominate, *vis solutiva* would be reinforced by the *substantial* life of Quinta Essentia. It would be as powerful in dissolving as the other in coagulating, resulting in a right equilibrium, just as Quinta Essentia produces normally. Such is *Hollandus's* opinion which he expounds from time to time in detail.

To this someone might object: Why then did all those who possessed veram Quintam Essentiam, including Hollandus himself, die? Why did they not always stay *in temperatum* so that no death could touch them? The answer is that God has set a goal for man's life, and Quinta Essentia cannot be effective against God's Almighty will. Just as at the time of the Flood the element water could predominate over the others at the behest of God and afterwards, although it went around the earth over the highest mountains by 15 ells (yards), it nevertheless had to suffer to be dried out again. That is why nothing can stop the Will of the Omnipresent Creator. At his command water must forget its power to quench, and fire, to burn, as the most glorious instruction may be found on this in the Book of Wisdom. So that the omnipresent, omnipotent Creator, when his hour has arrived to look after things and visit the sin, withholds all blessings, so that the very best medicine must also be of no avail. That is the reason why that fits in here what the author of the little tractate called *Mysterium Naturae Occultae* writes in these words: "As often as I think of the very serious threats which GOD the LORD holds before all trespassers of his Law, which I often do, I must get afraid to the utmost with my whole body and soul." The words of the Law are the following (Deut. 28. v.59): "The LORD will deal with you in a wonderful way, tormenting your seed, and it shall be great and special plagues and evil and special sicknesses." The same Prophet teaches that God's wrath will be so great on account of the sins that all salubrious medicines (which are the greatest gifts of God aside from the work of salvation) will be powerless.

How great now this curse is will all those see who experience in their bodies that all medicaments are cursed on account of their sins. That is why the *Medici* (doctors) did not wish to resort to the medicine but withdrew their support from diseases when they noticed something divine in them and were foiled as much by the variety as the multitude of the illnesses.

Aside from that, however, in order to speak more naturally about the much vaunted Quinta Essentia, the author just quoted duly says in the same little tractate: "One dose of this superb and very famous medicine rids man of all doubts, all "accidental" illnesses, renews the whole body, keeping it safe from all severe blows. For this spiritual medicine penetrates to the quick (literally: through mark and bones) to the root of the illness, and takes the lead in the weak nature by its manufactured, purifying power which improves it (the body) most beautifully in all ill health. It brings sleep for rest and appetite to eat. In truth, when a medicine cuts out the root of the sickness, also inducing sleep and appetite, I would not know what more it could do. And although the counter-chemists generally object to this that a saddle could not fit all horses nor a shoe do justice to all feet - with which usual argument they believe that they can cast destruction and doubt upon the unbelievable power of the universal medicine - the intelligent and rightminded will nevertheless clearly recognize what a great difference there is between a saddle and a spiritual *Medicina*. Then they must also see that a saddle may fit many thousands of horses and a shoe may be put on many thousands of feet. When now such a thing is conceded to and said about this medicine, I am satisfied; for I do not say that all men can hope to be helped by this medicine, but only many thousands.

In addition, some might object that a (single) thing could not agree with many. Then I ask from experience whether this one sun, air, fire, etc., is not good for many, yea, probably all. Our medicine is sun, fire, air, and spirit, which, if Master Prig does not understand, I consider it due to his ignorance and not to the art, just as the author's words state when he subsequently also gives good guidance for the preparation of that medicine out of Mercurial fount of metals.

My whole extensive presentation is meant to achieve that you give credit to the good teachings, drawn from Divine Light, of the highly experienced, lovable *Hollandus* and other philosophers, among whom I would count especially Jakob Boehme's *Teutonici Philosophi* writings which, aside from the Holy Scripture, are unequalled and put before you the heart and innermost center of all things. Do not think that it is due to emotion that I give such great honorary titles to such talented men, but rather (I am doing it) for your best, also to induce you to read without tiring, continually, with impartial attention, the books of the wise men. If you are sincere toward God and your neighbor, you will not regret it.

As an example you have Count *Bernhardus Trevisanus,* *Zacharius,* the French nobleman, and others who came to the most secret jewel of the philosophical stone without oral revelation, solely through the diligent reading of the books of other philosophers. It may well take somewhat long till some understanding is derived from reading; but one must not tire continuing until a right idea follows (suggest: one must not stop until a right understanding of the work follows), even if it were to happen only after several years. What are useful books for this purpose I have in part specified in my *Lucerna Salis Philosophorum.* Remember also to be on guard against Sophist books, which include those of the presently very famous Chemical Scribe G. As I understand it, they have caused my words to be misunderstood by many, as though I tried to diminish him in the minds of the fanciers, which is not my intention at all. Instead, I wish to set forth (my views) as I see them, without respect of persons.

Accordingly, all philosophers consider it Sophistry to look for something in a place where it is not, or to establish some processes against the orderly course of nature, which also he, through whom all things are made does not allow this highest philosophical teaching to do. Is it possible to gather grapes from thorns or figs from thistles? I therefore sought in my above-mentioned (work) to call the attention of the fanciers of the blessed philosophical work to the fact that all those philosophical processes of the said author, whether they have in parts been customary, or in parts newly invented by him, are to be considered Sophistry in regard to their application to the work. Reason: The age-old philosophers did not know anything about such things, did not distill, calcinate, sublimate, imbibe, etc., in such a way; but if they wished to propagate a thing, they simply took its specific seed, each according to its kind and species, and put it into its own *matrix,* thus processing it according to nature, as the propagation of a human being, a kernel or a plant shows. And since the philosophical *opus regeneration* is precisely based upon the simple course of nature, is quite easy and simple in itself, yet is the greatest gift of God in this world next to the soul's salvation, the philosophers have described it in veiled language, so that those who might have bad intentions would thereby be led astray, while the rightly-motivated, by praying for divine help, could snatch the truth from them (or "cull the truth"). That is how all present-day chemical disciples will speak, for their works are quite sophistic according to the letter and are needs written in such a way that those who intend to misuse them are misled thereby. But they are nevertheless truly philosophical and can lead to wisdom and its treasures.

Geber also used such a style in a masterly fashion. He himself says that he wrote in that way so that his opinion should not remain secret for the intelligent; that the mediocre should find it hard enough to understand him, whereas the ignorant should be miserably excluded from the sense. That is also why Count *Bernhardus* complains bitterly about *Geber,* as do *Archelaus, Ratin, Rupescilla* and others, that they are mostly mixed with Sophistic processes which he, Bernhardus, fought very much in the beginning. Not that he wished to disgrace the famous philosophers of Sophistry by his work, for it is a masterful work in itself, as may be seen by that which the esteemed author of *Vere Veritatis* writes, as printed in the "Wasserstein der Weisen" by Dr. Adam von Bodenstein, page 259, (namely) that he was a Master of Sophistry and had written many books on Alchemy, full of Sophist rules. They had looked as if a powerful understanding was contained in them, while they were nevertheless without foundation. However, he did not wish to burden his person with that work and was now saying that he had been a *Medicus* and a good Sophist. But according to his own writings, he had not well understood the natural art of Alchemy or *Secreta Saturnae.*

Therefore he wished to tell the simple people who base themselves upon such things, to guard against them, because their temptation is great but their truth bad.

Accordingly, I wanted to suggest with my book that every seeker should tread carefully, not lightly working according to the letter of some process. Rather, he should at first work with praying and reflecting about whether he could really start (the process) and whether he had actually understood the author's meaning, before losing labor, expense and time in vain in it. But I do not intend to diminish such authors or anyone else with saying so. For as little as the literal content of all such processes with all their rules will be found by every reader without any mistakes, just as little should everything be destroyed on account of one or several mistakes. Rather, one should stop short of all speculations which might incite a man to more inventions and knowledge. And I cannot say anything else but that much good is contained in the work of the said author, especially in his P.O., one piece of which pays enough for the price it costs.

The very precious *Secretum Philosophorum,* however, should not be sought in that kind of processes, and should much less be clearly described each time.

Unless someone wished willingly to draw upon his head the curses of the philosophers, as when *Ratis* says that if anyone were to divulge this supreme good to someone unworthy of it, he would become a violator (desecrator) and breaker of the divine secret.

Raimund Lullus says: "He shall be condemned at Judgment Day." The author of the great *Rosarium* says: "He shall be cursed and die of a stroke." *Basilius Valentinus* announces: "Such a one could not be reconciled to God, and would fall to the devil everywhere." *Hollandus* says from time to time: "He shall be purged temporarily and eternally." I, the disciple of those men who rest in God, say Amen to this, knowing that it shall be so. For if God deems a little faith of the size of a mustard seed, so worthy that it can move mountains, what will he do for the strong, magic faith of those men, as he sufficiently proved for Joshua by letting the sun stand still. Whether my unemotional opinion is received badly or well by one or another, can give or take little or nothing from me. Although I should prefer to see, and would yet desire, that no one should find cause for adversity in my plain explanation. Such a one would only harm himself in his inner man.

Instead, everyone who as a Christian gets totally rid of grudge, anger and enmity, also in regard to those who had offended him in the extreme, and who now becomes reconciled with his counter-value as if he had never been injured, as if it were all past, so that not the smallest spark of annoyance or displeasure stays in his mind because of the offense; as long as a person does not act thus just as long as he does not find real Grace with God but says the prayer of the LORD for his own damnation; since the Heavenly Father, in Christ's words, will deal with us in the same way that we have dealt with our debtor. This all the more since he gave us an example in this and became reconciled with us through the slaughter of his own son, while it was not he who had offended us in the extreme, but we had offended him in the extreme. Enough of that for this time.

Now it remains for me to say — if someone did not know it yet - what *Isacus Hollandus* intended to do by writing such *opera* as Animalia, Vegetabilia, and Mineralia. The meaning of it is that all sublunary things, originating in the elements, are comprised in three different realms. Among animal (things) are included all animals, worms, birds, fish moved by life, stirring, and endowed with sensitivity, and everything requiring breath, in addition to all substances coming from and out of them. The vegetable (things) comprise everything that grows and greens, out of the earth, leaves, grass, wood, stalks, blades and what there is about them in roots, fruit and other matters. Likewise some such greening vegetable (things) coming from the water, such as duckweed, etc. By mineral (things) we understand all the things that are coagulated within the earth, such as the ore of all metals and minerals, likewise various rocks, and whatever mountain juices, Sulphur, alum, etc., are brought or boiled out of the earth.

The noblest subjects of these three realms are the human being, wine and gold, which are greatly interrelated, as the philosophers long ago discovered. They also taught how to prepare the animal stone from man, the vegetable stone from wine, and the mineral stone from gold-nature, or *altero Solis,* which three stones contain all the power of all nature within themselves, especially the last one on account of its powerful projection, and they are a truly divine mystery as they (the philosophers) proclaim unanimously (as I intend to describe in future in detail in my *Harmonia,* please it God).

With all this *Hollandus* deals in his writings in full detail, and he has compiled the *magnalia* and secret operations of each realm in a special opus .... going from one to another and explaining one through another. Which I greatly desire the reader to understand well; and closing with this, I am bringing him under the wings of Grace of the (heavenly God) recommending him with all my heart for the very necessary understanding.

- Finis -

A Word from the Publisher

Thank you for purchasing this small work from The R.A.M.S. Library of Alchemy. During his lifetime, Hans Nintzel was dedicated to the identification, acquisition, study, retyping and, when necessary, translation of what he considered to be the most important known works on Alchemy. Hans was assisted by his sparse network of fellow Alchemists, all members of the Restorers of Alchemical Manuscripts Society (R.A.M.S.). I was an active member of R.A.M.S.

My goal is to publish all of the works originally made available through R.A.M.S. as photocopies. To facilitate this, I have chosen to have the books professionally printed. I also have a few titles that I intend to add to the original R.A.M.S. Library, selected by strict criteria established by Hans.

If you have a work on Alchemy that you believe should be a part of the R.A.M.S. Library, please contact me through R.A.M.S. Publishing Company.

Philip N. Wheeler